A Brief History of Digital Electronics

By
Doug Domke

Along with Some Interesting Personal Stories, Observations and Predictions

A History of Digital Electronics

All Text Copyright © 2019
By Doug Domke
All Rights Reserved
Second Edition, V 1.1

Printed through Lulu.com
ISBN: 978-1-387-80104-6

Contents

About the Author

Doug Domke spent his career in the semiconductor / computer-chip industry. He is also an avid computer hobbyist.   He has a Bachelor of Science degree in Physics. He is currently retired and lives in Phoenix, Arizona.

In addition to "A History of Digital Electronics", Doug is also the author of:

"A Small Book on Physics" (2012)

Each chapter is only a few pages, but gives you an overview of an entire subject. One chapter tells you all about "black holes". Another tells you what "quantum mechanics" is. You will also learn about the more classical areas of physics. What is "thermodynamics" or the "physics of sound"? You'll learn what physicists still don't understand, where they are in their quest, and what's likely to happen next! Other topics include: The Big Bang, Gravity before and after Einstein, String Theory, The size and age of the Universe, Electromagnetic Waves, Solid State Physics, Lasers and Holograms, as well as Sub-atomic Particles.

"What's Inside the Black Box?" (2014)

Technology is all around us today, and for most of us, the inner workings are incomprehensible. What you might not know is that most of the inner workings are incomprehensible to even the people who designed them. That's because modern technology is the result of collective intelligence, where one person may make a significant contribution, but doesn't know and doesn't need to know how the complete system works. We get great technology by compartmentalizing complex technological challenges. One guy figures out how a certain input generates a specific output. Once that's done, everyone can use it. The inner workings can remain unknown to everyone else. It's what's known as a "black box". In this book we will examine a number of items you use every day, like your microwave oven, digital camera or automatic transmission. We'll take them apart and see how they work. Then you will know "What's Inside the Black Box"!

"What Is Empty Space?" (2016)

Why ask what empty space really is? It is the nature of space itself that seems to be at the heart of the mysteries in physics and cosmology today. Albert Einstein told us that space is a something. He showed it could be stretched and warped when he published General Relativity in 1915. But over the last one hundred years, the nature of empty space has continued to become more mysterious and confusing. We've learned a lot about the Universe in the last one hundred years. We've even learned a lot about the nature of empty space. But we still have a lot more to learn! That is the subject of this book.

Preface

I have been thinking about writing this book for the past three years, without doing anything more than creating outlines, and perhaps a little research. One problem has been the scope. I started with the idea of writing the history of the personal computer. It begins with some history of the semiconductor industry, continues with the emergence of microprocessors during the 1970's, and evolves through PCs, Macs, and eventually smart phones.

But the subject is enormous, and much been written about it already. There are also big areas of the subject that I currently know little about. I would have to do extensive research, and even with a resulting 600 page book, I could add little to what has already been written.

And there's another issue with writing a history of the personal computer. This is an area where I have a lot of firsthand knowledge, and I want to write about my personal experiences. I'd like to write a book that's only half history and half my own stories.

So I am narrowing the scope some, focusing on the digital electronics revolution that took place in the second half of the 20th century. I am not attempting to write a comprehensive history of the personal computer, or even a full history of the emergence of digital electronics. I will get into some history of the semiconductor industry in which I worked in my entire career. I will also talk about the personal computer in some detail, but not in as much depth as I would have if this were a complete history of the personal computer. I will also try to tell a few personal stories along the way. So this book is roughly 50% factual history and another 50% personal stories. Hopefully in this way, I can give the reader an understanding of how the digital electronics evolution unfolded, both by explaining it in terms of history and by relating some personal experiences that relate to that evolution.

Another issue I have struggled with in writing this book is how technical to make the discussion. I want to write a book for everyone, not just semiconductor experts or personal computer hobbyists. But at the same time, you can't talk about these technologies and the industries they created without talking about the technology that makes it all possible.

So as you read this book, you will see me all over the place trying to deal with this issue. Sometimes I get way too technical for the reader with no background knowledge. If I start talking about something you don't understand, don't worry. Just skip over it and we'll soon move on to something else.

At other times, I will gloss over or skip details that someone familiar with these technologies will think is critically important. To those I apologize. It is simply difficult to find and maintain a proper balance. So I have let myself go back and forth on this, sometimes getting way too detailed and sometimes trying too hard to keep it simple.

I have included in this book a lot of extra stuff that some people may find interesting. It's a little more technical than the rest of the book, so these bonus chapters may not be for everyone. I labelled them *"If You're Interested"*. They are, in some cases, things I have written in the past for other purposes, but they seem sufficiently relevant to include again here. I've tried to place them where they seem relevant to the previous chapter. These chapters have been printed in italics to help show they are extra reading for those who are interested. They can be skipped by anyone who isn't interested.

I've called this book "A Brief History of Digital Electronics", but the last two chapters are definitely not history. There are some fascinating changes taking place in computer technology right now, and I couldn't resist some speculation about where all this is going.

Doug Domke
October 16, 2019

Acknowledgements: Special thanks to my my friend Bill Altonen and my stepdaughter Ginger Sainz for their help in proofreading and editing this book.

Life Before Personal Computers and the Internet

If personal computers, smartphones and the Internet have been around all your life, which is true for everyone younger than 35 years old today in 2019, it is probably hard to imagine how the world functioned before we had these things.   In fact, even for people like me, who watched all this stuff come along over the second half of the 20th century, it's hard to remember how things got done without the Internet, computers, and digital communications. They have changed everything!

What was it like being born in 1880 and seeing cars and airplanes come along and change everything during your lifetime? It was probably about the same as it was for me - being born in 1944 and seeing digital electronics change almost every aspect of life!

So this chapter is a brief look at the before and after. The idea is to illustrate how much this stuff we take for granted today has changed the world. These changes have all taken place in the last 35 years, and yet, we quickly get accustomed to them, and soon forget what it was like without them. So in this chapter we will remember and compare.

Let's start with phones. When I was young, we had what were called "party lines", as in 2-party or 4-party lines.   One phone line was shared between 2 or 4 families. With the preferred 2 party line, you only heard the phone ring when it was for your family.   (In case you're wondering, the phone call was carried on a pair of wires. The ring signal could be placed between one wire and ground for one family, and between the other wire and ground for the second family.) The 4 party line was not as desirable. In addition to 4 families having to share one phone line, it used a long ring for one family and two short rings for another family. In either case, it was not uncommon to pick up the phone and find another family using your phone line. You had to wait for the other family to finish before you could use the phone. Obviously, privacy was non-existent. And, needless to say, long phone calls were considered somewhat rude in that environment.

Up until 1970, we used rotary dial phones. The clicks they made as the dial rotated told the system what number you were calling. The touch-tone keypad system (which used dual frequency tones to transmit numbers) replaced the dial telephone around 1970, and by then most families had a single-party line – no more sharing the phone line with other families.

Mobile phones – mainly built into a car – started appearing in the 70's as well, but they were very rare – basically a toy for the very wealthy. They weren't cell phones. There was no "cell phone system". There were a few frequencies reserved for mobile phones, and operators manually connected these mobile phones to land-line phones. They were open lines, meaning anyone with a mobile phone could listen in on other people's calls by just selecting a frequency that was in use. They were not particularly reliable – there was no cell tower infrastructure. Service was pretty much limited to big cities, and disconnects were common.

A mobile phone before cell phones

Cellphones began appearing in the 1980s, but they were big, heavy, and expensive. They were a status symbol – definitely not for everyone.

By the 1990s, however, everyone was getting cell phones, and by 2000, it seems like every teenager had their own cell phone. How quickly things had changed. Today we take for granted the idea that every person has their own phone, their own private line, their own personal phone number.

Let's not forget about long distance. Today, most cell phone plans come with unlimited long distance service included in a package. But not very long ago, long distance phone calls were an expensive luxury. Writing letters was inexpensive, and the preferred way to communicate. Long distance phone calls were generally prohibitively expensive. You might call your parents who lived in another state once every couple of months. And it would cost you $10 or $20 every time you did!

Now let's change subjects and look at the Internet. It traces its beginnings back to the 1970s with a primitive network called ARPANET that allowed universities to exchange files and share research documents. ARPANET gradually grew for the next 20 years to become the Internet, but it still remained primarily the realm of universities and governments until the World Wide Web emerged in the early 1990s. The "web" utilized HTML formatted files linked together with "hyperlinks". It was a system of HTML coded files, web-browsers to view them on, and servers to deliver them, all riding on the existing Internet network. The worldwide web made an enormous difference. You no longer needed to know UNIX to use the Internet. Suddenly anyone and everyone could use the Internet!

Between 1990 and today, the Internet has become the primary way information is disseminated around the World! It is so pervasive today that even those of us who were around well before it struggle to remember how we survived without it!

Here are some examples of life before the Internet.

Mail Order Shopping

You had to mail in a request to get a catalog. When you got your catalog, you filled out an order form and mailed in your order. You typically got an order confirmation by first class mail within a week, and then waited another week or two for what you ordered to actually arrive. The system worked, but it was slow. Choices and price competition were limited.

Research of Any Kind

There was the library of course. The city had one, the university had one. We had one at work. They are still there today, but they were the main source of information before the Internet!

Didn't want to go to the library? Then perhaps you had a set of encyclopedias at home. They were a frustrating combination: a huge set of books in which there was very little detailed information on any particular subject. And frequently no information at all on a subject!

Correspondence

Before email via Internet became commonplace around 1990, the main way to correspond was regular mail. If you were in a hurry, you paid extra for airmail. And if you were in a really big hurry, you picked up the phone and called.

The environment for correspondence at work was more interesting. Writing something and copying dozens of people like we do today with email simply wasn't an option. If you wanted to write something, you wrote it out by hand and gave it to a secretary. She would type it and give it back for changes, corrections, etc. She would then make copies of the final version, and they would be delivered to the recipients by an internal mail delivery department once or twice per day. Correspondence between different company locations would have to use regular mail.

In general, we spent far less time communicating in writing. It wasn't a very efficient process, so it was only used when it was important to document something.

Planning a Vacation

Now here is something that has really changed. Today I do everything on the Internet: research where I want to go, make my airline, hotel, and car rental reservations, find and print maps, find points of interest and how to get to them. How on Earth did I do all this before the Internet?

Travel brochures weren't just nice, they
were pretty much essential before the Internet

Travel agents were one answer. Research at the library or in travel magazines was another. Obviously, we had fewer choices, less information, and less competition for your vacation dollar. We always managed even without the Internet, but the Internet was a huge improvement!

As a result of cell phones and the Internet, our whole concept of being connected is fundamentally different in recent years. Today we assume constant, uninterrupted communication and connectivity as a way of life. Even 30 years ago, it was not unusual to go on vacation or a business trip, and be out of contact for a week or two, except perhaps for a postcard mailed home. The fact that we could telephone in case of an emergency made us feel that we were connected in a way people had never experienced in the past. But we didn't expect to stay in touch every day.

We were grateful for telephones and airmail, because only 150 years ago, a simple trip of more than a few hundred miles put you completely out of touch with your family for months at a time.

But now, no matter where we go, we can reach for our cell phone or tablet, and be in touch with anyone, or find any information we want. We take this all for granted, but these capabilities have only been with us a very short time!

Photography

What has changed as much as photography? Cameras use to be bulky, and you didn't carry one around unless 1) you were on vacation, or 2) you were a professional photographer. Film was expensive, developing and printing even more expensive. Want instant results? Polaroid pictures were even more expensive. We used cameras sparingly and took our time to get the right picture.

Today everyone has a very good camera built into their phone. Everyone takes their phone/camera everywhere. Any interesting event will be photographed by numerous bystanders. Taking pictures in unlimited quantities is basically free once you buy your smartphone. Parents take hundreds or even thousands of pictures of their kids every month, because "why not", it's basically free!

The quality of photos afforded by digital electronics is vastly improved as well. Auto focus, auto-exposure, color correction, ease of editing, etc. make almost everyone a pretty good photographer. And then there's video! Every smartphone now comes with the ability to take HD video. Aside from the issues associated with getting the lighting and sound quality right, your phone can take high-definition, broadcast quality television video!

Instructions

Here is one more interesting example. Products of all kinds, especially ones you had to assemble yourself, used to come with really good instructions. They had to or no one would ever figure out how to put them together or use them. Today, the written instructions that come with anything are terrible. Why? Because we have the Internet, smart phone apps, YouTube, and a myriad of other ways to get help. Written instructions have almost disappeared, frequently being nothing more than website reference to where you can find the real instructions.

Life Has Changed!

There are life changing effects from digital electronics everywhere you look. We could go on all day. When is the last time you did any of these things? If you're young enough, you likely never did most of these things!

- Memorized a phone number or used a phone book to find a company to do work around your house?

- Solved a simple math calculation in your head?

- Bought a CD or had a record collection?

- Made a photo album?

- Called a theater to get movie times?

- Looked up the spelling of a word in the dictionary?

- Watched shows when they are broadcast live? (Even when I watch something live, I record it and start 10 minutes late, so I can fast forward through the commercials.)

- Sent a handwritten letter?

- Carried cash?

- Used a travel agent?

- Got your old checks back from the bank every month, or even wrote a check at all for that matter?

Early History of Digital Electronics

Where do we start our history of digital electronics? Most of what we will talk about in this book has its origins in the semiconductor industry – the amazing progress we have made in the last 50 years cramming more and more transistors onto a single piece of silicon. But to be fair, many brilliant men and women made great contributions to computers and digital electronics even before the invention of the transistor. So in this chapter we will look at some of the pioneers who got this all started.

Jacquard's Programmable Loom

Long before computers, people started using punched cards and punched tapes to store digital information. In 1804,

Joseph-Marie Jacquard developed a loom in which the pattern being woven was controlled by a paper tape. The paper tape could be changed without changing the mechanical design of the loom. This was a landmark achievement in programmability.

Today we think of computers as electronic by necessity, but computers were mechanical in the beginning. An abacus is an example of a really primitive mechanical computer. But a machine built in 1833 by Charles Babbage, an English mechanical engineer, is generally regarded as the first computer. Called the Analytical Engine, it was completely mechanical, but contained a arithmetic logic unit, control flow in the form of conditional branching and loops, and integrated memory, making it the first design for a general-purpose computer,

Charles Babbage's Analytic Engine

Babbage used punched cards for input and output. The analytic engine employed ordinary base-10 fixed-point arithmetic.

Another early computer pioneer was Alan Turing. In 1936, he wrote a paper defining the requirements for a general purpose computing machine - what is today referred to as a universal Turing machine. He essentially proved that any problem that could be described in an algorithm could be solved by this general purpose machine. He also introduced the idea that such a machine could perform the tasks by executing a program stored on tape, allowing the machine to be programmable. Turing is generally regarded as the first great contributor to computer science and artificial intelligence as we know them today.

Allen Turing

Another early computer pioneer was John von Neumann. In 1945, Von Neumann took Turing's work and went a step further. He defined the structure of a digital computer that met

Turing's criteria for a complete general purpose computer. The machine would have the following components:

- A processing unit that contains an arithmetic logic unit and processor registers

- A control unit that contains an instruction register and program counter

- Memory that stores data and instructions

- External mass storage

- Input and output mechanisms

John Von Neumann

Computers built to follow Von Neumann's criteria are said to comply with the Von Neumann architecture. Almost all computers in existence today use Von Neumann architecture!

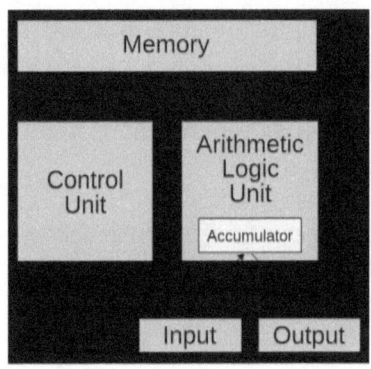
Design of the Von Neumann Architecture

In the 1940's, there was suddenly a lot of interest in computers. The subject was evolving very quickly from an academic one into a desire for a tangible product. Computers were desired for a very specific reason - World War II was going on and computers could help win wars. A big part of World War II was coded transmissions and code-breaking was an important science. Computers were seen as hugely important tools for both code-breaking and other warfare technologies. Consequently, governments were funding the development of electronic digital computers.

The ENIAC Computer

The ENIAC (Electronic Numerical Integrator and Computer) pictured on the previous page was the first electronic general-purpose computer. Completed in 1945, it performed ballistics trajectory calculations for the United States Army. Of course like all early computers built before semiconductors, it was built with vacuum tubes, which meant huge physical size, cost and power consumption. Specifically, ENIAC contained 20,000 vacuum tubes, weighed 27 tons, occupied 1800 sq. ft., and consumed 150 kilowatts of electricity.

Another noteworthy early computer was the first IBM computer - IBM 701 Electronic Data Processing Machine was introduced in April 1952.

IBM 701 processor frame, showing 1071 of the vacuum tubes

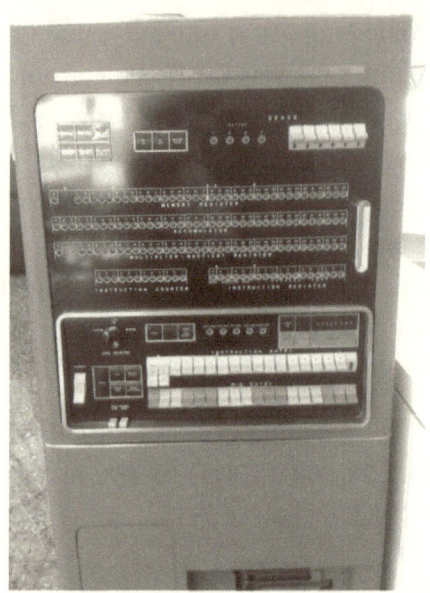

IBM 701 Front Panel

One interesting thing about the IBM 701 was its use of cathode ray tubes (like old TV picture tubes) as memory storage.

The "Williams Tubes" shown above could each store 1024 bits of memory. The IBM 701 used 72 of these tubes as its main processing memory. They were eventually replaced by magnetic core memory.

Above is a picture of an IBM 701 logic board with its eight vacuum tubes. The computer needed hundreds of modules like this. As mentioned already, constructing computers with vacuum tubes made them physically huge and very expensive. So even before the invention of the transistor, computer manufacturers were busy trying to miniaturize the hardware.

The photo on the previous page shows the shrinking of computer hardware. It shows three different generations of the same vacuum tube logic board and finally a transistorized version. The transistor version is made with discrete transistors, so even it is huge compared with putting the whole thing on a tiny chip of silicon as was later done with an integrated circuit.

The Early History of Semiconductors

As we mentioned in the last chapter, the first computers were built with vacuum tubes in the 1940s and 1950s. They filled entire buildings, cost millions of dollars, and consumed incredible amounts of power!

But digital electronics really started to become a reality with the emergence of semiconductors, so we need to look at what they are, and how they came into existence. Semiconductors get their name because they conduct electricity, but not as well as metals such as copper. How well they conduct electricity, and the methods with which they conduct it are the subjects of solid state physics. I have included a couple chapters after this one, if you wish to learn more on the basics of semiconductors.

William Shockley, Walter Brattain, John Bardeen, inventors of the transistor. They won the Nobel Prize for Physics in 1956 for their accomplishment.

Semiconductors became the key technology driving digital electronics with the invention of the transistor at Bell Labs in 1948. Like the vacuum tube, the transistor could be used as a switch to control the flow of electricity, but even from its inception, the transistor had huge benefits compared with the vacuum tube. It was much smaller and used much less power. And as we shall talk about later in much more detail, improvements in transistor technology made it possible to manufacture smaller and smaller transistors, which consumed less and less power. This has made it possible to put that early computer that filled a whole building on one tiny chip of silicon. In fact, todays microprocessors are systems tens of thousands of times more complex and powerful than those early computers, and yet, they fit on a single ½ inch square piece of silicon.

Following the initial invention of the transistor in 1948, it didn't take long for the importance of the transistor to be realized. Numerous companies sprang up in the 1950s to manufacture and sell transistors. Western Electric began manufacturing transistors in 1951. Texas Instruments followed about a year later. The semiconductor material in use in the early 1950s was germanium. The silicon technology we are all familiar with today came later. The first transistors were known as point contact germanium transistors. They were big, very fragile, and were not an instant success. But they gave engineers and electronics companies something to work with.

One of the most interesting applications for these new transistors, at least from the point of view of consumers, was the transistor radio. There was a lot of immediate appeal, because portable radios had become very popular in the 1950s. They were made possible by the manufacture of very small vacuum tubes. While they were portable, they were somewhat bulky. More importantly, they went through a lot of batteries. Two D-cell flashlight batteries powered the filaments of the tubes, and an expensive 67 volt battery supplied the tubes high-voltage plate current.

A portable Philco vacuum tube AM radio

Transistors promised to make portable radios much smaller and bring their power consumption way down. And most importantly, they would make the expense for batteries much more reasonable!

Many companies in both the U.S. and Japan were eager to get their hands on inexpensive, mass produced transistors suitable for manufacturing radios. It was a key factor in creating an early market demand for transistors.

An early transistor radio

Germanium junction transistors soon replaced the original point contract transistors. Junction transistors were more rugged, more reliable, easier to mass produce, and had better characteristics than point contact transistors. They were the devices radio manufacturers needed to make radios.

The first germanium junction transistor available to the public was Raytheon's PNP transistor, the CK722. The CK722 was originally priced at $7.60 each, but had dropped in price by 1956 to 99 cents.

Germanium transistors have limitations. They can't operate at temperatures higher than 85°C (185°F), and generally have high leakage currents. Silicon is a better material. It can operate at temperatures as high as 200°C (392°F), and has leakage currents several orders of magnitude lower than germanium. So by the early 1950s, companies were already trying to make silicon based transistors. The first significant success came from Gordon Teal at Texas Instruments in 1954. He is credited with making the first commercially practical silicon transistor. But manufacturing silicon transistors in high volume and at low prices took a few more years to materialize.

In 1955, William Shockley, who had invented the transistor seven years earlier, left Bell Labs in New Jersey, and started up Shockley Semiconductor in Palo Alto, California. Shockley was able to put together a very talented team of young scientists. While Shockley Semiconductor only lasted a few years, it was important for two reasons. First, as the original semiconductor company in the Bay Area, it gave birth to Silicon Valley, the area south of San Francisco which still today is the epicenter of semiconductor, computer, and Internet technologies. Second, those talented young scientists went on to be founders of many of the successful semiconductor companies which still exist today. Sixty-five different semiconductor companies can trace their roots back to Shockley Semiconductor.

Eight young scientists at Shockley left to start a new company - Fairchild Semiconductor, with financial backing from

Fairchild Camera and Instruments. Two of these eight were Robert Noyce and Gordon Moore, who later founded Intel.

At Fairchild, another of the eight founders, Jean Hoerni, invented a new process to manufacture transistors – the planar process. In this method of manufacturing, transistors are basically printed onto a wafer of silicon using photographic technology. Then silicon dioxide is used to mask the silicon, while impurities are diffused into the silicon to form the various regions of the transistor. Fairchild's planar process was a monumental breakthrough for many reasons. It allowed transistors to be truly mass produced for the first time. They became cheaper and more reliable. The planar process made most other ways for making transistors obsolete. The planar process also made it possible to put multiple transistors on single silicon chip and wire them together. It is the process that facilitated the birth of integrated circuits, and the one still in use today to make them.

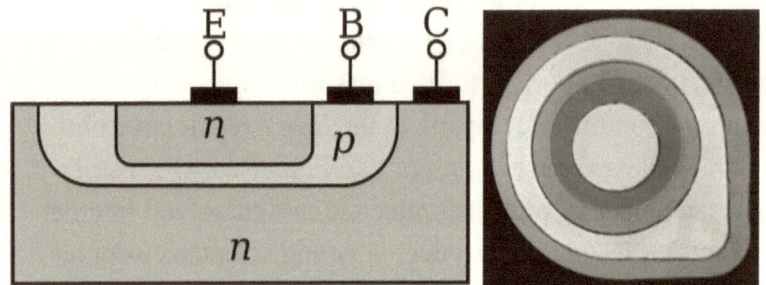

A side-view drawing (left) and top-view photo (right) of an early NPN planar transistor with the Emitter, Base, and Collector terminals.

The planar process, along with many other innovations such as epitaxial layer transistors (which give transistors lower on-resistance and improved high-frequency performance), put Fairchild into a position of industry leader during the late 50s and early 60's.

Out in Phoenix, Arizona, another company was making a name for itself. Motorola first established a semiconductor laboratory in 1955, and was soon manufacturing germanium power transistors. Motorola had a very specific interest in transistors. It was the world's largest producer of car radios, which like everything else were made with vacuum tubes in 1950. But vacuum tubes put a heavy load on the car's battery. They also were fairly unreliable, especially when bounced around all day in a car. So Motorola wanted to build transistor car radios. Their big incentive for getting into the semiconductor business in the mid-50s was that no one could make power transistors capable of powering the speakers of a car radio. So Motorola decided to make power transistors, and later to make all kinds of other power semiconductor devices. Motorola soon became a major player in all aspects of the semiconductor business and one of its most successful competitors.

Motorola made many great technological contributions to the semiconductor industry over the years. I will mention one of them here for two reasons. First, it is a natural extension of the planar technology pioneered at Fairchild that we have already discussed, and second, it was a contribution made by a good friend of mine, Jack Haenichen.

Making high voltage PNP planar transistors had proven surprisingly difficult. To get high voltage, you had to start with a lightly doped p-type wafer. The silicon oxide on the surface of the device caused a thin inversion layer to form so that the surface of the P-type collector region acted like N-type silicon, causing the collector region to effectively short out. Jack added a more highly doped P type ring around the collector, which cut off the inversion layer and allowed the whole industry to now make high voltage planar PNP transistors.

Meanwhile, back at Fairchild and also at Texas Instruments, the planar process was creating the beginning of the integrated circuit industry in the early 1960s. Robert Noyce at Fairchild and Jack Kilby of Texas Instruments were both experimenting with putting multiple transistors on a single chip and wiring them together with deposited aluminum wires.

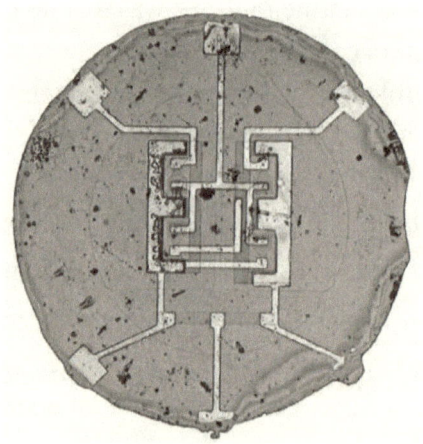

Fairchild's first integrated circuit, a 4 transistor,
5 resistor flip-flop circuit introduced in 1960

So the integrated circuit was born in the early 60s, and continued to evolve and mature throughout the 1960s. We need to look at some of the important technologies that helped.

So for, we have only talked about bipolar junction transistors. These are basically current amplifiers, where a small amount of current flowing into (in the case of an NPN transistor) the base of the device controls a much larger amount of current flowing into the collector. In the 1950 and 60s, most commercial transistors were these bipolar junction types.

However another type of transistor had been the subject of lab work and experiments throughout this period, the field effect transistor or FET. This device uses an electric field or voltage applied to its gate to control the flow of current from its source to its drain.

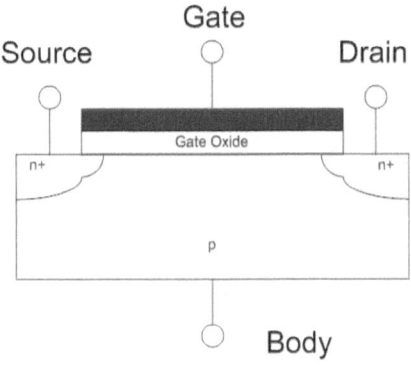

A particular version of the FET uses a metal gate insulated from the silicon beneath it by a layer of silicon dioxide. Shown above, this FET is known as a MOSFET, where MOS stands for metal, oxide, silicon. It turns out that MOSFET transistors are superior to bipolar junction transistors for building integrated circuits. They have a very simple structure which makes them small and ideal for putting lots of transistor elements on one small piece of silicon.

So during the 60s, companies were developing simple integrated circuits using bipolar junction technology, but as the decade progressed and companies started designing larger and more complex integrated devices, MOSFET technology started to be used. One especially attractive application for large integrated circuit technology was computer memory. Magnetic core memory, or even semiconductor memory built with discrete transistors, required huge amounts of space and power, so creating large amounts of small, low power memory was a big priority for the new integrated circuit industry.

In 1968, Robert Noyce and Gordon Moore left Fairchild to found the first company dedicated purely to integrated circuits. That was Intel. Its first product was a computer memory chip using MOSFET technology. The 1101 chip stored 256 bits of data. It was the first semiconductor static RAM (random access memory).

Then in 1970, Intel introduced the 1103, with 1024 bits of RAM on a single chip. To get that many bits on a single chip, they switched to dynamic RAM, which only require a single transistor for each bit stored. It's called dynamic because it can only store its data for a few milliseconds before it needs to be refreshed, but even with the nuisance for refreshing, the density or number of bits able to be stored on a single chip makes dynamic RAM the way to go. The 1103 was a big success for Intel, and turned out to be the beginning of the end for large, expensive magnetic core memory. In fact, dynamic RAM is still used as the main memory in every computer today.

Putting 1024 bits or 1 Kbits of memory on a single silicon chip was a big accomplishment back in 1970. But by 1973, Intel and other companies were putting 4 Kbits on a single chip. By 1976, they were putting 16 Kbits on a single chip. And by 1980, they were putting 64 Kbits on a single chip.

Gordon Moore

The growth in the size of memory chips was just getting started in the 1970s. Today in 2019, we can put 2 GBytes or 16 GBits on a single chip. That's 16 million times what we could be stored on a single chip in 1970! This exponential growth in the number of transistors that can be placed on a single silicon chip was actually predicted by Gordon Moore while he was still at Fairchild. Moore predicted in 1965 that the number of transistors that could be put on a single silicon chip could be expected to double every two years.

Microprocessor transistor counts 1971-2011 & Moore's law

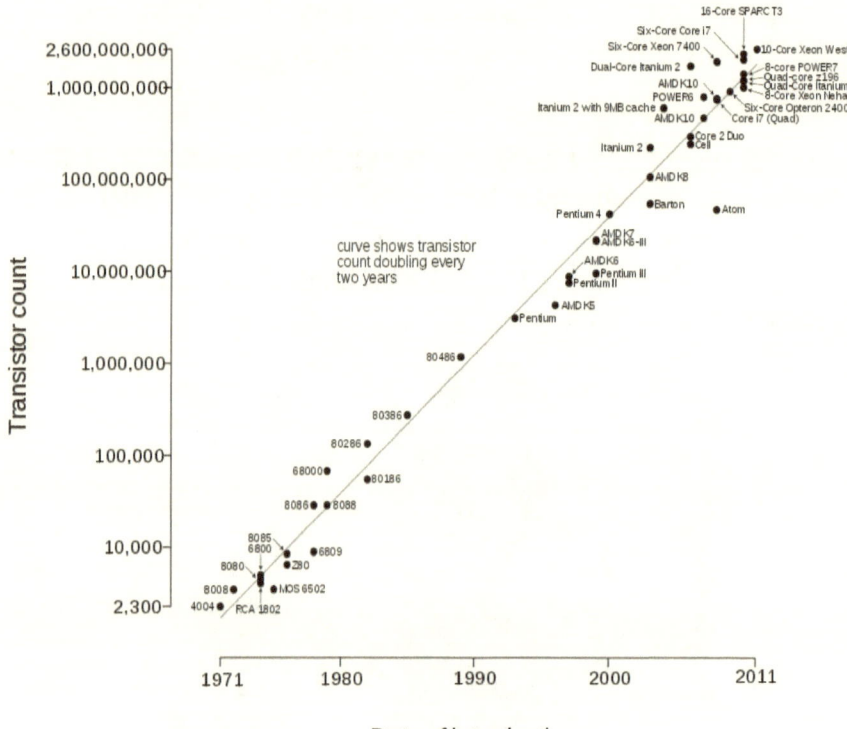

This graph illustrates Moore's Law and
how it has held up over the 40 years

This somewhat daring prediction became known as Moore's Law. Ironically, it wasn't daring at all, but rather somewhat conservative. The transistor count has actually been doubling about every 18 months, and, incredibly, it has continued at that pace not just through the 70s, but for the past 50 years!

This concludes our early history of the semiconductor industry. When we continue the history in a later chapter, we'll be ready to talk about the complex integrated circuits that have changed everything: calculator chips, microprocessors, and the other systems on a single chip of silicon.

If You're Interested:
Solid State Physics

You have probably heard of solid state physics, perhaps in conjunction with semiconductors and computer chips. It is the science of solid materials and in particular crystalline materials where the atoms are tightly packed in an orderly repeating pattern.

Silicon crystal is one such material, and it is the raw material for the majority of semiconductor devices - transistors and integrated circuits that make today's electronics possible. We will discuss the physics of these materials (including semiconductors) in this chapter. Then, in the next chapter, "Semiconductor Devices", we will explain how solid state physics leads to the electrical function of diodes and transistors that make up modern electronics.

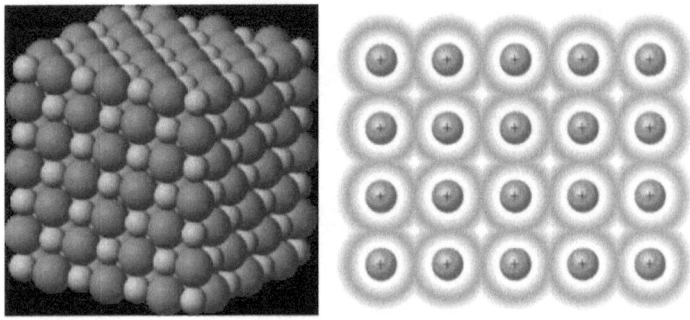

On the left is a sodium chloride crystal (chloride is the larger, darker atoms), showing the orderly, tightly packed nature of a crystalline material. At the right is a depiction of the positive nuclei in a crystal surrounded by the swarm of electrons that are shared within the material.

Solid materials have certain properties that are a function of the material as a whole, rather than being a function of their individual atoms. Conduction of electricity is one of these properties and is of particular interest in solid state physics. Conduction of electricity is a function of how tightly the electrons in the material are bound to its individual atoms.

For an insulator, the electrons are all tightly bound to individual atoms, and cannot flow through the material. In a metal, many of the electrons are completely unbound and free to flow at ease throughout the material.

The differences in electrical conductivity of materials are derived from the energy structure of the crystalline material. The spacing between atoms and how the electrons are bound and shared between atoms effects what energy states are available for electrons.

There are some allowed energy bands and some other forbidden energy bands known as "band gaps". In the lowest allowed energy band, called the valance band, the electrons are trapped within a single atom. There is then an energy band gap followed by one or more higher-energy bands where electrons are free to flow between atoms; these are called the conduction bands. The larger the band gap, the harder it is for electrons to move freely, and the more the material acts like an insulator.

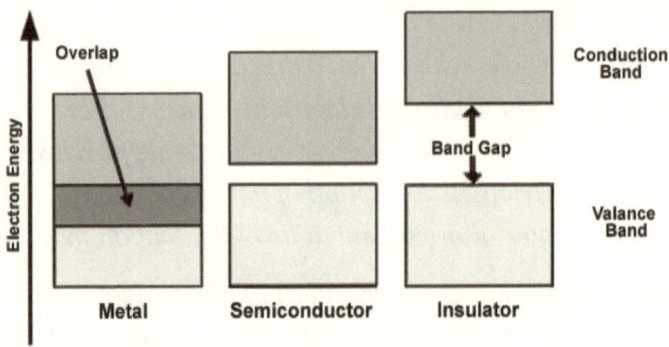

In metals, the band gap doesn't really exist – the valance band and conduction bands overlap each other, allowing electrons to flow freely. In an insulator, the band gap is very large, making it nearly impossible for electrons to move through the material. In between, as the name implies, are semiconductors. We will spend the rest of this chapter exploring silicone as a semiconductor.

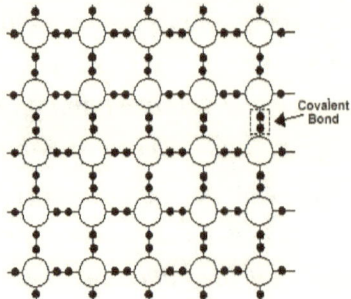

Semiconductors like silicon are poor conductors in their pure "intrinsic" state. The atoms within the crystal are bound together with covalent bonds, meaning that the electrons are shared between two atoms and a very large band gap prevents conduction. For all practical purposes, pure or intrinsic silicon is an insulator. But what makes semiconductors

special is that small amounts of impurities in them are capable of making them into pretty good conductors. Two common impurities used in silicon semiconductor technology are phosphorous and boron. We will now look at why these impurities turn silicon from an insulator into a conductor of electricity.

First we have silicon with its covalent bonds and no free electrons. But when we add a small amount of phosphorus to the crystal structure, phosphorus has one extra electron compared to silicon. That extra electron, called a donor electron, easily has enough energy to pass the band gap and becomes a conduction electron, freely flowing through the silicon. Add more phosphorus and we get more free electrons. Silicon becomes a better and better conductor as we add small amounts of phosphorus. Phosphorus and the other elements that add donor electrons to silicon are called N-dopants and silicon with an N-dopant is called N-type silicon. (The N just stands for negative, since the donor electrons are negative.)

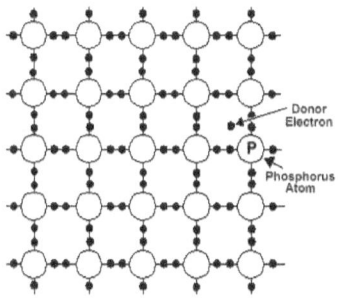

There is another way, however, to make silicon conductive. Other dopants like boron can be added to the silicon crystal. Unlike phosphorus, which has an extra electron, boron is one electron short. It leaves a hole in the covalent bond structure, and it turns out this hole is almost as mobile as a free electron. A "hole" is just the absence of an electron, but for all practical purposes, moving holes behave just like a positive charged particle moving through the silicon. Holes conduct electricity. Boron and the other elements which generate holes in silicon are called P-dopants and silicon with a P-dopant is called P-type silicon. (The P just stands for positive, since holes act as positive changes.)

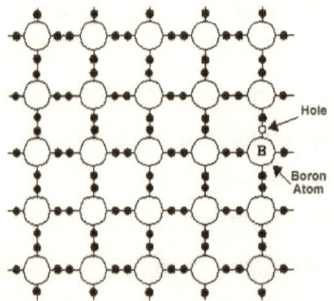

So we have just seen how semiconductors can be made to conduct electricity. How well they conduct can be controlled using dopants like phosphorus and boron. In the next chapter, we will see how this technology is applied in practice to make semiconductor components, like diodes, transistors and integrated circuits.

If You're Interested:
Semiconductor Devices

Below are a few examples of semiconductor devices, the components of modern electronic products. The transistor and the light emitting diode (LED) are discrete components, individual electronic elements. The integrated circuit and the microprocessor (which is a particular kind of integrated circuit) have an entire circuit, made from numerous discrete elements, wired together on a single chip. All of these devices, except for the LED, are made of silicon.

Microprocessor   Transistor   Integrated Circuit   Light Emitting Diode

In this chapter, we will look at how the properties of semiconductors that we saw in the last chapter are used to switch, amplify, or otherwise control electric currents in today's electronic devices. Some understanding of electricity is useful here, but not essential.

Let's start with a small piece of silicon, which is doped on one side with an N-type dopant and on the other side with a P-type dopant. This structure is called a diode.

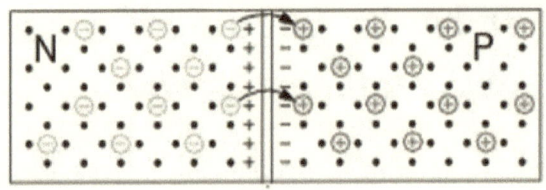

Electrons near the NP junction diffuse across the junction and combine with holes. The result is twofold: 1) a region forms around the junction without holes or electrons, a so-called depletion region with no electric carriers 2) because of the electrons moving across the junction, we have a positive charge left on the N side of the junction and a negative charge on the P side of the junction. In silicon, the electric field produced by this charge is roughly 0.7 volts.

Now let's connect a battery to this device and try to pass an electric current through it. If we apply a negative voltage to the N-silicon and a positive voltage to the P-silicon, as soon as the voltage exceeds the 0.7 volt junction barrier voltage, current will begin to flow as electrons move toward the junction on one side and holes move toward the junction on the other. As they combine, the process continues and current flows. We say that the diode and its junction are forward biased.

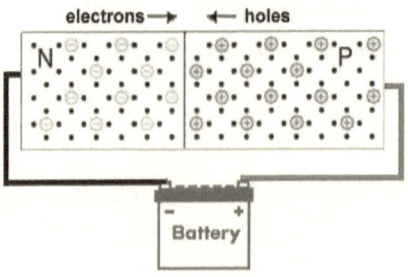

If we now apply a positive voltage to the N-silicon and a negative voltage to the P-silicon, we will pull both electrons and holes away from the junction, increasing the width of the depletion region. No current is allowed to flow. We say in this case that the diode and its junction are reversed biased.

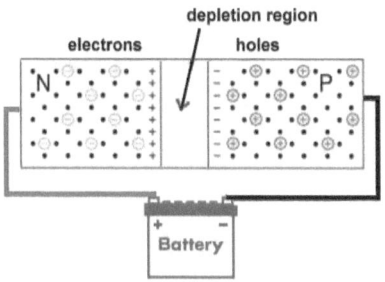

Consequently, our diode passes current in one direction, but blocks current flow in the other direction. This is the definition of a diode!

Now let's complicate things slightly by having two junctions and a sandwich made up of two slices of N-silicon with a layer of P-silicon in between. This structure is called an NPN transistor.

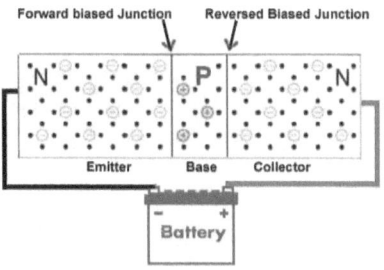

When we apply a voltage across our transistor, one junction is forward biased and can conduct current, but the second one is reversed bias, so no current is able to flow. The N-silicon on

the forward biased junction is called the emitter. The N-silicon on the reverse biased junction is called the collector. The P-silicon in the middle is called the base.

So now let's make a third connection, this time to the base of our transistor. That allows us to pass a small current through the forward biased junction, bypassing the reversed biased one. This causes electrons to flow across the emitter-base junction.

A surprise, however, is that, unlike holes, electrons in the base region can quite easily flow across the second, reverse biased junction to be "collected" in the N-silicon collector. In fact, most electrons crossing the first junction end up being collected by the collector rather than simply becoming base current. The result is that a rather small base current produces a rather large collector current. Small changes in base current result in large changes in collector current.

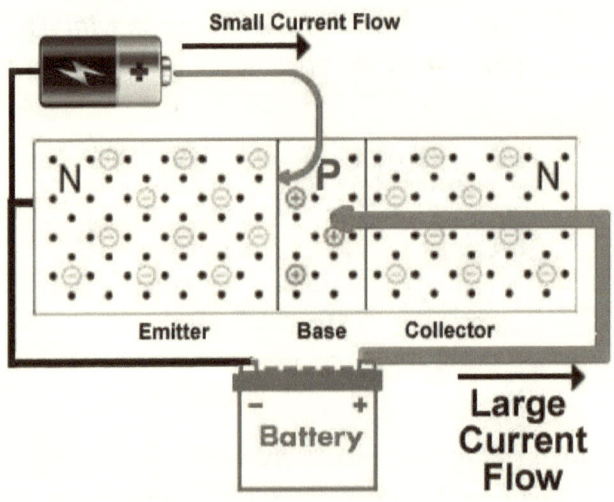

So our transistor can be used as a switch, turning large amounts of current on or off using a small current. It can also be used as an amplifier, using a small signal input at the base to create a large signal output at the collector.

This chapter gives you a quick look at how diodes and bipolar transistors work. There are in fact many different kinds of transistors. Today's silicon integrated circuits contain thousands and even millions of transistors on a single chip. Most integrated circuits made today from silicon use a technology called metal-oxide semiconductor (MOS) field effect transistors (FETs). They are the technology of choice because of a combination of good speed, small size, and low power dissipation.

My Introduction
to Digital Electronics

As mentioned in a previous chapter, the first transistor that was available to the general public was Raytheon's CK722. And as it turns out, I was one of those hobbyists who couldn't wait to get my hands on a few CK722 transistors. I began experimenting with the CK722 in 1956 as a 12 year old boy, making a two transistor amplifier as my first project. (My dad, an electrical engineer, had helped me build a two vacuum tube amplifier two years earlier.)

But an amplifier isn't digital. My exposure to digital electronics came gradually over the next several years. In this chapter, I will give you some personal history, including my early exposure to computers. We'll look some more at how technology has changed over the past 70 years.

I tinkered with all sorts of electronics projects during junior high school and high school. When I was in 8th grade, selenium photocells were relatively new, at least as an inexpensive electronic component, so I entered the San Francisco Area Bay Area Science Fair with a project showing what could be done with photocells. I included a burglar alarm and a sun powered radio among other things. My science fair entry was one of about 30 that were invited to participate in IEEE's Future Engineers Show later that same year.

I'm not telling you all this because my project was that great. But at the IEEE Future Engineer's show I was really inspired by what some of the older kids there had done. The project that impressed me the most was a maze solving computer.

You used switches on the control panel to create a maze, and then this machine would find its way out of that maze. It would also remember how it got out, so if you asked to try again, it went directly there by the most direct path. If you tricked it by changing the maze, it would first try what had worked before, but as soon as that failed, it would go back into its search mode and find a new way out. The heart of this machine was a big rotary stepping relay. It was surplus telephone switching equipment, and it had 100 positions. As I recall, it was spiral shaped and at least a foot tall! It was very impressive, and the whole thing really piqued my fascination with computers.

Here is a 25 positon telephone stepping relay.
The 100 position one was much bigger and wrapped
around in a spiral. I couldn't find a picture of one,
so they must have been rare, even back in the late 1950s.

My first actual hands-on experience with digital electronics was a slightly bizarre project in high school. Somebody who knew I was an electronics hobbyist ended up with several boxes of surplus electronics. One item among them was a bunch of DPDT (double-pole, double throw) mechanical relays. He offered them to me for free. I then had to figure out what, if anything, I wanted to do with them. I found a rather simple circuit with which a flip-flop could be fabricated from two DPDT relays. (A flip-flop is a standard and commonplace digital circuit element.) I ended up building a sixteen stage binary counter using 32 relays, with a row of neon lights displaying the output of the counter. At maximum speed, it could run at about 300 counts per minute, and with all 32 relays clicking away at the same time, it created a terrible racket when running. But it was a real conversation piece. I ended up donating it to my high school physics class when I graduated.

It sounds crazy today, but I graduated from college in 1969 with a degree in Physics without ever using a computer. Colleges had mainframe computers, and there were classes where you could learn to program them. But unless you were majoring in computer sciences, there wasn't much need to use computers. Everything I did in Physics was calculated to 3 place accuracy on a circular slide rule.

Actually, I starting using a circular slide rule in high school, like the one above. By the time I graduated from college, it was like an extension of my body. The biggest problem with a slide rule was keeping track of where the decimal point is – you had to do that part in your head, while going through a long series of calculations. After maybe 10 calculations you would get to your final answer and end up with 437. But was that 43.7 or 0.437 or 4.37×10^{-8}?

Let's move on and talk about computers! I first learned about actually using computers when my friend Don Rose, a math major at Cal Berkeley, took a computer programming course. He was learning the programming language Fortran. This was probably 1964. He had to write his program by hand, and then use a "keypunch" machine to input the program into a series of punch cards, like the one below. Each card held a single line of program instructions.

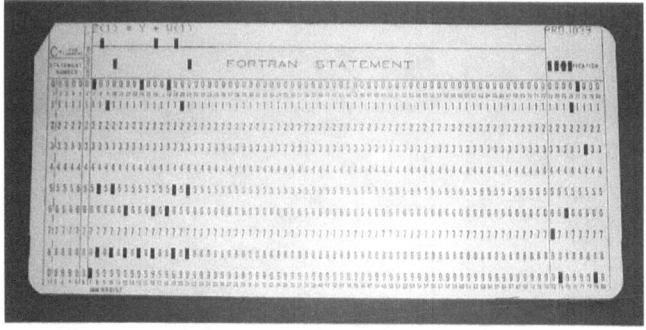

Once you had your program written and punched into a stack of punch cards, you took the stack over to the Berkeley Computer Center and dropped it off. Everything ran in "batch mode" then, so you didn't know exactly when your program would execute. You had to come back 24 hours later, and pick up a piece of paper with your program's results. Naturally, like any programmer learning a new language, you had some errors, and had to fix them and try again. Each iteration added another 24 hours to the cycle, so debugging a program could take forever. It seemed like an excruciating process to me at the time.

My own hands-on encounter with a computer began when time-sharing got popular in the early 70s. By then, I was working for Motorola Semiconductor. I had access to their Xerox Sigma 9 system and learned to program it in Basic. Unlike my friend Don's experience at Berkeley, by then computers had become interactive and accessible in real time. I could type my program into a portable terminal, type RUN, and get a result a minute or two later. I started out having to use a teletype machine to communicate with the Sigma 9, but soon the Texas Instrument Silent 700 portable terminal came out. It was a big improvement!

Teletype Terminal   TI Silent 700 Portable Terminal

I started developing simple applications to help summarize and display data, and to consolidate the large amounts of data we were collecting. After a while, I got pretty good at it. I really enjoyed programming - wrote a few game programs in addition to the serious stuff. What appealed to me was that, unlike most of the rest of life, mastery over a computer - getting it to do what you wanted - had nothing to do with good or bad luck. It was just patience and hard work!

I never did take a course in computer programming or any course related to computers. But I read a lot of books, worked hard, and gradually became very knowledgeable – never an expert in anything, but I had some knowledge in a lot of different areas.

Working at Fairchild

I went to work for Fairchild Semiconductor in Mountain View, California in late 1962, shortly after graduating from high school. I worked as a technician in the Materials Department, while going to college at night. We made the silicon wafers, from which transistors were fabricated. That meant growing cylindrical single crystals of ultra-pure silicon, slicing them into wafers, polishing them to a mirror finish, and for some wafers, depositing a layer of high-resistivity silicon on top of the wafer using epitaxial deposition.

One of the things that we had to do was measure the thickness of these epitaxial layers. We used a Beckman Interferometer, and to get the thickness you had to read the wavelength of two interference peaks from a graph the interferometer printed, and then go through this lengthy calculation on a slide rule to come up with the thickness in microns. This was an opportunity to put my slide rule talents to work. I came up with a set of 3 printed logarithm scales placed on a large sheet of paper in a way that allowed you to lay a straight edge across them. By placing the straight edge on the peak wavelengths on 2 of the scales, you could then read the thickness of the epitaxial layer directly in microns from where the straight edge crossed the third scale. This silly little tool saved hundreds of hours of labor and was still in use when I left.

When I first starting working at Fairchild, I worked in a small building a couple of miles from Fairchild's main plant in Mountain View, and we had our own resident accountant. He had a machine that I found fascinating, as I had never seen anything like it before.

It was a Marchant 8CM, a mechanical calculator capable of not only addition and subtraction, but multiplication and division as well. Ask it to divide a 6 digit number by a 4 digit number, and it began whirling and clicking that went on for perhaps as long as 2 minutes, but then up came the answer! Very impressive! The accountant was very proud of his machine, and quite pleased to show me how it worked and how to use it. I'm pretty sure I snuck into his office a few times, when he was out for lunch, just so I could play with it.

About 1½ years after joining Fairchild, our Materials department moved into Fairchild's main plant and headquarters on Whisman Road. At this point, all the famous founders of semiconductor companies were still working at Fairchild. I can't say I knew them; I was just a peon. But I saw them all in the cafeteria and knew who they were. Charles Sporck was there, and went on to found National Semiconductor. Robert Noyce and Gordon Moore were there, before founding Intel. Jerry Sanders was there, before founding AMD Semiconductor. It was hard to miss Jerry; he was our very flamboyant head of marketing, and I think everyone knew who he was.

One interesting side note about Fairchild is that it was then right down the street from the somewhat famous Wagon Wheel bar, which was not only the hangout for Fairchild employees after work on Friday's, it was kind of the meeting grounds for the whole semiconductor industry. Many of the high level managers showed up there on Friday nights, and it was an opportunity for peons like me to meet and talk with the managers. (I wasn't 21 yet, but no one seemed to notice.)

I was very fortunate to experience a special time at Fairchild. With hindsight, I think I was too young to really appreciate that a huge new industry and a technological revolution were being born right before my eyes.

I was only 20 years old, so everyone around me seemed much older, but in reality, the whole workforce was very young. Most of the managers were in their late 20s or thirties, and even the senior management was in their early 40's. The whole place was very casual and there didn't seem to be a lot of rules. It must have given the human resources managers a lot of sleepless nights. We had two big parties a year: one in the middle of summer, and one at Christmas. The entire company was invited and there was an open bar that went for many hours. The week following each of these parties was very interesting. Rumors were flying everywhere about all the wild things that had happened. Several divorces seemed to result from each party!

Another thing about Fairchild, which is a little off the subject, but nevertheless fascinating, was the state of industrial safety in the 1960s. We used silene, a clear liquid compound of silicon and hydrogen, to grow epitaxial layers on silicon wafers. It was extremely toxic, but worse yet, it spontaneously burned when exposed to the oxygen in the atmosphere. We used to handle this stuff in large, easily breakable quartz flasks. If one started leaking, we would put on gas masks. Then we would run down the halls of the plant, and throw it out the back door into the parking lot, where it exploded in a big ball of white smoke. Fortunately, no one ever got hurt that way to my knowledge.

However, we also used 1500 or 2000 watt RF generators to grow our epitaxial layers. These generators were capable of seriously injuring people if they got exposed to the RF output directly. RF energy travels along the surface of your body rather than internally, so it doesn't electrocute you, but it can give you terrible skin burns. We had technicians like me who maintained this equipment, but women actually ran them, loading in wafers, using the RF to heat the wafers white hot and then depositing a layer of silicon on them. The RF generator obviously had to be off during the loading and unloading of the wafers, but there were no interlocks to prevent you from forgetting to turn them off and reaching in while the RF generator was on. So every few months, one of the women operators got burned, and every couple of years, someone ended up in the hospital. Every time it happened there were meetings and discussions about the need to fix the problem, but it never did get fixed in the 3½ years I was there.

Here is one last observation about Fairchild. There were many semiconductor companies in the area around Mountain View in the early 60s. That meant there was lots of competition to hire people with any experience and technical expertise in semiconductors. People changed companies frequently. I couldn't believe the turnover at Fairchild. Every other company around was trying to hire people from Fairchild. By the time I left after 3½ years, there was only one person left in the Materials Department, out of maybe 50 engineers, managers, and technicians who had been there longer than me!

More History of Semiconductors

By 1970, the semiconductor industry was making real integrated circuits. By that I mean we were making single chips of silicon with up to 1000 or more transistor elements. The integrated circuit business wasn't huge yet. Discrete components like small-signal transistors, power transistors, diodes, rectifiers, etc. still made up 90% of the industry. But everyone could see that integrated circuits were going to be a huge market. So every semiconductor company was getting into the integrated circuit business and some new companies were making integrated circuits exclusively.

By the late 60s, electronic desktop calculators had started to appear, but they were big and very expensive.

 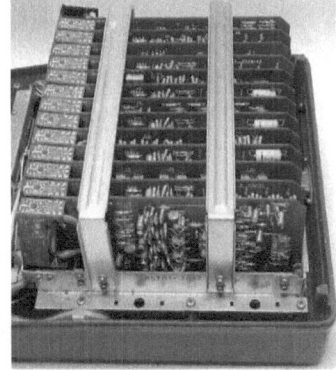

The Canon Canola 130 in 1968 contained 13 circuit boards full of components. It cost $1000.

So even before 1970, a major goal of integrated circuit manufacturers was to integrate the electronics of a desktop calculator on a single chip of silicon. In late 1970, Mostek, an integrated circuit start-up in Texas and only two years old, produced the first calculator on a chip for Busicom, a Japanese desktop calculator manufacturer.

Only a few months later in 1971, Texas Instruments introduced their first calculator chip. Unlike Mostek, TI had bigger plans than just making silicon chips for desktop calculator companies. TI jumped into the calculator business itself, and did so in a big way. Hewlett-Packard jumped into the calculator business soon after that, and for the next few years the "calculator wars" raged, with everyone getting into the business of either calculator chips, calculators, or both.

In 1971, the 6000 transistors on a chip required to supply basic calculator functions was pushing the limits of what could be put on a single chip. But let's not forget Moore's Law. Three years later in 1974, it was possible to put 15,000 transistors on a single chip of silicon. So the "calculator wars" were fought on two fronts: those who could produce calculators for the lowest cost, and those who could produce calculators with the most features.

The first handheld calculators added, subtracted, multiplied and divided. That was it! But soon single chips provided scientific calculators, financial calculators, and other specialty calculators with all kinds of special functions and features.

The first calculators built with a single chip still cost about $300, but by the end of the decade, you could buy a calculator for $3! The power of integrated circuits and the digital revolution were on full display.

Something else, however, even bigger than the calculator, started to happen in the semiconductor industry during the late 60s and early 70's. Computers were still very expensive, and still required massive numbers of components, but they were now fabricated from semiconductors. Integrated circuit manufacturers had been busy in the late 60s trying to get the heart of the computer, the central processing unit (CPU), integrated onto a few chips of silicon. In the process, they were learning and acquiring a lot of computer expertise. So by 1971, Intel, TI, Motorola, and several other major semiconductor players were trying to put the CPU on a single chip. Integrated circuit technology in 1971 was finally capable of putting enough transistors on a single chip to make it possible.

So in 1971, Intel introduced the first microprocessor, a computer CPU on a single chip of silicon. The Intel 4004 was a 4 bit microprocessor with 2300 transistor elements on the chip. The 4 bit term means that the data paths within the chip are 4 bits wide, only able to handle numbers in a range from 0 to 15. That made it difficult to do anything really useful with this CPU, so it wasn't a big commercial success, but it was significant as a proof of concept. We now had a CPU on a single chip of silicon.

The Intel 4004

Several other companies introduced microprocessors over the next couple of years. Intel introduced an 8 bit version, the 8008. But as with the calculator, things changed rapidly between 1971 and 1974. By 1974, a second generation of microprocessors started to appear. These were high performance 8 bit processors, capable of serious applications in industrial automation or even as general purpose computers.

The two most successful 1974 entries were Intel's 8080 and Motorola's 6800. These two 8 bit microprocessors were what started the personal computer revolution! In1975, kits began to appear which allowed hobbyists to build their own computer. MITS introduced the Altair 8800, followed by the IMSAI 8080. Southwest Technical introduced the SWTPC6800, and Sphere Corporation came out with the Sphere 1, a fully assembled 6800 based system. The personal computer industry was off and running. We will discuss this in much more detail in later chapters.

To finish our discussion of semiconductor history in the 70s, obviously by 1975, everyone was starting to see a huge potential in personal computers. Numerous companies began to work on microprocessors, and there was enormous competition to make better, faster, more powerful, and cheaper microprocessors. One significant entry into the 8 bit market was the MOS Technologies' 6502, similar to the 6800 but with some minor improvements and a lower price. It was the processor selected for the Apple I and II. Another significant 8 bit processor was the Zilog Z80, which remained a very popular high performance processor for many years, even after most 8 bit processors had be replaced with 16 bit machines.

Finally, it was still the late 70s when 16 bit microprocessors started to emerge. Intel's 8086 and 8088 and Motorola's 68000 come to mind as the most successful of these. But again, we will talk about them more in a later chapter.

Motorola Semiconductor Part I

It is not clear where in this book I should talk about my years working for Motorola. I worked there from January 1966 until February 2001. Many things I could write about here should have come earlier in this book, while other things should have come later. In this chapter, I will talk about my early years at Motorola, and then in two other chapters, I'll talk about my later years at Motorola.

I am going to start my discussion of Motorola by mentioning someone who worked there during my first couple of years there: Wilf Corrigan. If you ever worked in the semiconductor industry, you probably knew the name. Wilf was at that time an operations manager, regarded as a very smart workaholic. He seemed to be working at all hours of the day and night. I didn't really know him except perhaps to exchange Hellos in the hallway, but everyone knew who he was. In 1968 he became one of eight senior managers at Motorola who all left at once to join Fairchild. The departure of most of the top management all at once had a major impact on Motorola Semiconductor, which took a year or two to fully recover from. But the main reason I bring up Wilf Corrigan is that he soon was running Fairchild Semiconductor. And then in 1978 he went on to be the founder of VSI Logic – yet another Fairchild employee who went on to found his own company!

Wilf Corrigan

I started at Motorola as a technician in Thyristors. I was hired by Jack Haenichen, the guy I talked about earlier who invented the annular ring that made high voltage PNP transistors possible. He and I have remained very close friends for many years.

Jack Haenichen, as he appeared in
Motorola's 1971 Annual Report

The culture at Motorola Semiconductor was similar in many ways to Fairchild. I felt comfortable there immediately because it felt familiar, like I was back at Fairchild. But there were some significant differences. At Motorola, people sometimes changed jobs or moved to a different department, but they hardly ever changed companies. Everybody seemed to be a loyal Motorolan and had been there several years. There was a family orientation where having your spouse or child working there too was encouraged. That was very different than Fairchild.

As I said, I worked in the Thyristor Operation. Thyristors are discrete semiconductor power switches. Small ones are used in products like light dimmers. Big ones are used in industrial applications like the robots that build cars. In addition to the power devices themselves, we designed and manufactured various triggering devices to turn on these power switches - exotic small signal devices like unijunction transistors and three layer diodes.

This was a very cool, special time in the history of the semiconductor industry. It hadn't gotten insanely high-tech yet. I was able to learn every step in the process of manufacturing semiconductors:

1. Making the silicon wafers like I did at Fairchild

2. Designing devices, meaning setting the dimensions of a device, junction depths, doping concentrations, etc.

3. Photolithography which involves making masks and using photo-projection equipment to print the device's design on silicon.

4. Diffusion (or later ion implantation) to drive impurities into various regions of the device.

5. Metallization where aluminum or gold is deposited and then cut to make connections to the device

6. Assembly where a) the silicon chip is bonded to a metal frame; b) wires are bonded to the terminals of the devices and brought out to external terminals; and c) the device is packaged in metal or molded plastic.

7. Testing where the finished device is checked out electrically to ensure that it actually functions per its design specifications.

In 1967, I was able to do everything myself. I could literally operate every piece of equipment required to perform every step in every process above.

That was a pretty good trick back in 1967, but the opportunity to do it at all quickly disappeared within the next few years.

By the mid-70's, semiconductor manufacturing was super high-tech. An engineer or technician was a photolithography engineer or an assembly engineer or a test engineer, or a device designer. Everything was highly specialized and it had become impossible to move between these highly specialized fields. So my time in Thyristors was a special opportunity for me to learn many facets of the business very quickly.

And since the industry was so young, opportunities for advancement came quickly as well. I was soon supervising several other technicians, and as soon as I got my degree in physics in 1969, I was immediately promoted to Product Manager, supervising both technicians and engineers. I was 25 years old.

I remember how old I was when I became a product manager for a very specific reason. Soon afterwards, Motorola Semiconductor's then general manager, Tom Connors, gave a speech complaining about the fact that his management team included a bunch of young, inexperienced 25 year old product managers. Unfortunately I was one of them.

Tom Connors

I remember saying in an earlier chapter that my first real experience with computer's was programming Motorola's time-sharing system, the Sigma 9, writing programs to summarize data. But I did have one earlier experience while working in Thyristors. One of the first places computers showed up in our business was automating test.

Before computers, testing was incredibly labor intensive. You would put a test operator in front of a curve tracer (a test instrument that could be set up to measure a specific characteristic of the device) with a bag of perhaps 250 parts. In the first pass, they would look at each part, examine several features and cull out the rejects. They would then reconfigure the curve tracer to examine some different characteristics, and repeat the process. The number of passes varied with the part, but after a couple of hours, they would return with a bag of good parts and a bag of rejects.

Needless to say, this process looked like a very attractive opportunity for computer automation. In our test facility, we finally got our hands on an automatic test system. It consisted of an automatic parts feeder/sorter, a bunch of electronics to 1) supply power and inputs to the device being tested, 2) make measurements of the device being tested, and finally, a DEC PDP-8 minicomputer to control everything. You programmed it on a teletype machine by specifying for each test: the voltage and/or current on some pins, what to measure on other pins, and finally measurement limits for what constitutes a good part. Then you created another test, and another until the part is fully tested. The resulting test program was saved by printing it out on a punched paper tape. It was saved that way until you were ready to test that particular device, and then it could be read back into the computer from the punched paper tape.

This automated approach to testing semiconductor parts produced a huge increase in productivity. One operator running an automated test system could test 5000 parts per hour, compared to perhaps 100 parts per hour for pure manual test.

We talked briefly in the last chapter about calculator chips. I bought my first calculator in 1971. This was for home, not for work at Motorola. They were expensive, but I had to have one. It was the exact model shown above.

It was definitely not cheap! I think it cost about $250, which is like $1500 today. And it didn't do anything except add, subtract, multiply, and divide. And when the display overflowed, you just got an error – no scientific notation here. But it was absolutely creepy how fast the price of a calculator came down. As we said in the last chapter, by 1975, you could buy one for $10. This is really a great example of the price we pay for being an early adopter.

Personal Computer Pioneer??

As mentioned in a previous chapter, the first high-performance microprocessor chips were introduced in 1974, and not long after that, kits started emerging for building a primitive home computer. I was really fascinated by all this. But of course, I had a fairly demanding full-time job at Motorola, so this was just another hobby for me.

It was 1975 when the first home computer kit came out. It was the Altair 8800. I didn't buy one, but I read all about it. You had to program it one byte at a time. You literally selected an address with one bank of switches, set the 8 bits going into that address on another bank of switches, and hit the load button. The address advanced by one, and you could then program the switches for the next byte. Needless to say, just getting it to flash the front panel lights in an entertaining pattern was an accomplishment you could be proud of.

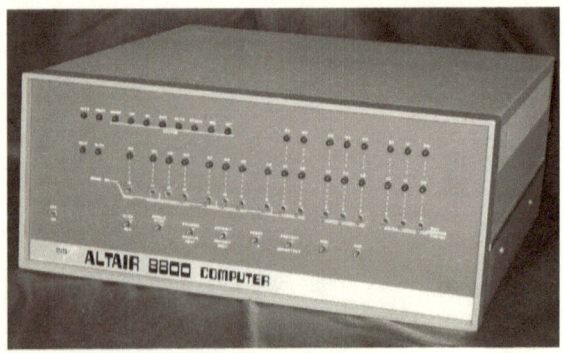

Altair 8800 Computer

By late 1975, there were other kits. A buddy of mine named Tom Pappas was a computer programmer, and he bought a Sphere Computer. Tom bought it preassembled, rather than as a kit. Unlike the Altair 8800, it had a built-in primitive monitor, so you could do a lot more with it. He actually let me borrow it and play with it for a few days.

My big accomplishment with Tom's Sphere was creating a simulation of a bouncing ball on the screen of the monitor. Since that required simulating gravity, Tom was very impressed, and wondered how I did it. There were no high level languages, no math functions, nothing . . . I basically did it by cheating! I calculated by hand where the ball should be at each 1/10 of a second. Then I put a little table into my program with those numbers. The program looked up the number and used it to place the ball at the appropriate height to make it look like it was bouncing.

Sphere 1 Computer

One of the kits that came out in late '75 was the Southwest Technical PC6800, which used Motorola's 6800 microprocessor. Since I worked for Motorola, I didn't want to buy an Intel based kit. But I took the plunge shortly after the PC6800 kit came out and bought the kit.

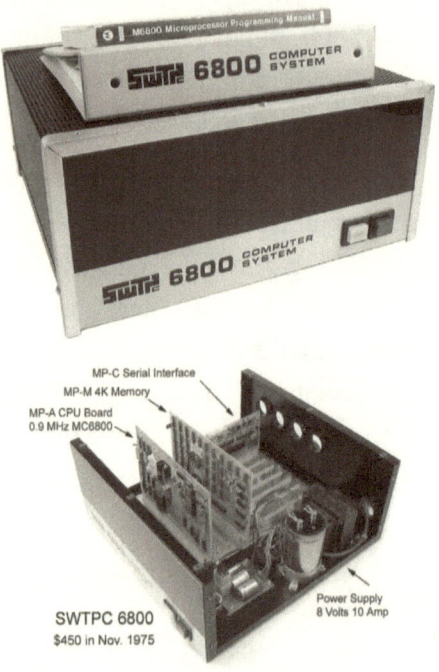

MP-C Serial Interface
MP-M 4K Memory
MP-A CPU Board
0.9 MHz MC6800

Power Supply
8 Volts 10 Amp

SWTPC 6800
$450 in Nov. 1975

Once I mounted and soldered the parts on all the boards and assembled the computer, I was able to run some simple tests to verify that it was working. It wasn't! I spent the next week trying to figure out why it wasn't working. Finally, I discovered a microscopic metal shaving shorting out 2 address lines on the main circuit board. Once I found it and removed it, my computer started working immediately.

However, you can't do too much with the computer by itself. Unlike the Altair, the SW Technical computer didn't even have a front panel. So my next project was to build a CRT terminal, which was another SW Technical kit.

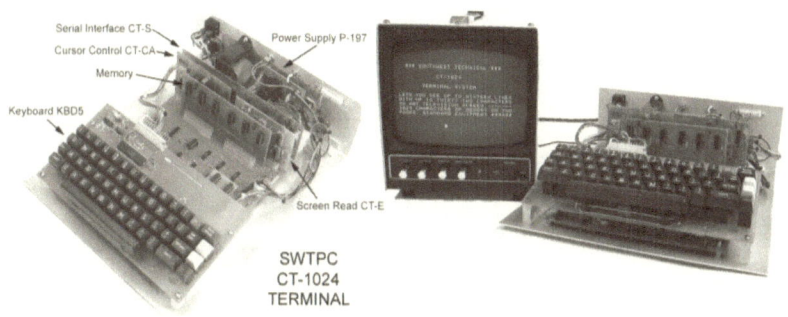

Serial Interface CT-S
Cursor Control CT-CA
Memory
Power Supply P-197
Keyboard KBD5
Screen Read CT-E

SWTPC
CT-1024
TERMINAL

This was a very privative CRT terminal. It displayed 32 characters per line and 16 lines. You had to supply your own TV as the output device, and most inexpensive TVs weren't fast enough or didn't have high enough bandwidth to display the characters crisply. So you had to be content as long as you could read the text. This CRT terminal was actually a bigger project than the computer itself, culminating in having to find the video detector inside the TV, and connect the output of the CRT kit into the TV at that point, in place of the picture signal coming from the tuner of the TV.

With these 2 kits completed and a TV attached, I was finally able to see the results of my efforts after about a month of working every evening. Although I still had to compile simple programs by hand, and load in the result one byte at a time, it was a lot easier to do it on a keyboard, rather than from switches on a front panel. There was no assembly language compiler, no high level languages, and no way to store a program when you finished it, but I actually had a working computer!

Next I started on a long series of projects that spanned the next couple of years. They added up to a whole lot of work. Here are those projects:

1) I thought the CRT terminal kit left a lot to be desired. The worst thing was that when you reached the bottom of the screen, the text didn't scroll up. You had to clear the screen by pushing a couple of keys on the keyboard, the screen went blank, and you were free to start typing again. But everything you had already typed and any responses from the computer were now gone!

With a degree in Physics and no formal education in electronics, I wasn't much of a circuit designer, but I finally invented my own system for getting the page to start scrolling when you reached the bottom of the page. It wasn't complicated, but it did require cutting a couple of runs on the circuit board, adding an IC on a new small circuit board, and wiring it into the existing system. I sent off my modifications to SW Technical, and they immediately modified their own design to match my changes. They also sent my modification plans to all existing owners of the CRT kit.

2) I had to have some way of storing programs. The computer had a serial output port, and it was fairly easy to write a program to output the program code. The problem was what to do with it. There were several simple schemes for building an audio modem that stored digital data as audio that could be recorded on an audio cassette tape player. SW Technical eventually came out with a kit for the modem, but I had already built my own by then. It had to be very carefully tuned

to get it working, and even when it worked it was slow and not entirely reliable.

I want to elaborate slightly on what I mean by "not entirely reliable." We take the accuracy of stored data today for granted. You can't tolerate bit errors buried in your programs or data. Computers can easily detect errors using a simple "check sum". You add up all the bytes you store, letting the total overflow. When you are done, you get an eight bit or 16 bit sum. When you read your data back, you check to see if the "checksum" you get when reading matched the one from when you stored the data. If it matches, then you have a reliable storage and retrieval system. I built all this into my simple cassette storage system, and it worked. But I sometimes had to store a program two or three times before I got a valid store. And even then I might occasionally get a "checksum" error when loading. So it was "not entirely reliable." But it did finally give me a way to store programs.

3) While it was possible to write very simple short programs by hand, i.e. looking up the code for each instruction, and literally compiling it by hand on a piece of paper, it was very slow and tedious. Because I worked at Motorola, I had access to a 6800 assembler that ran on the Sigma 9 time sharing system. By either staying late at work, or bringing home a portable CRT terminal with dial-up capability, I was able to write more complicated programs on the Sigma 9, and then print out the resulting code. I still had to type it into my computer by hand to get my resulting program to run, but it worked. And of course, once it was in and working, I could now save it on an audio cassette.

4) I wrote a simple program that demonstrated artificial intelligence. I got the idea from a book on computer games. The program was called Animals, and it tried to guess what animal you were thinking of. It started out with only a single question: "Does it live on a farm?" If you answered yes, it guessed a cow. If you answered no, it guessed a cat. Whenever you informed it that its guess was wrong, it asked what question it could have asked to distinguish between the animal you were thinking of and the animal it guessed. It saved all the questions, and gradually got smarter and smarter at guessing animals. But my computer started out with only 4000 bytes of RAM. The limiting factor for my Animals program was always memory. It would run out after about 10 minutes of playing Animals, and I would get an error message. I gave my program to SW Technical, as all of us in those days were dying to get our hands on anything useful or entertaining that could run on these early computers. Southwest Technical thought it was great, thanked me profusely, and started offering it as a product, for sale on a cassette tape for about $10.

5) My next project was a big one. I didn't feel comfortable about using time on Motorola's Sigma 9 for my personal hobby. And it wasn't nearly as convenient as it would be to have an assembler actually running on my own computer. The problem was that no one had a small, 6800 native assembler. Motorola had one, but it took a lot more memory and additional hardware than we had on these hobbyist systems. So I decided to write my own. Of course, a project this size had to be built on the Sigma 9, but I could also tell myself that this would be the last time I had to use the Sigma 9. It took

many weeks to get the assembler designed, programmed, and actually working, but I did eventually get there.

Once I started using it, I starting finding programming bugs. Also, I kept wanting to add new features, so about once a week, I would make a bunch of changes and assemble a new version. While I had originally envisioned this as a very simple, primitive assembler, by the time I was finished, it pretty much had all the functions of a commercial product. I thought about trying to sell it, but realized I would end up spending way too much time supporting it, so that never happened. But it worked great. I went on using it myself for the next two or three years.

Bill Gates   Steve Jobs

Steve Wozniak   Me
(As you can see, I wasn't the only one back then with BIG glasses!)

To put all this in a broader context, while I was doing my thing, a guy named Bill Gates was dropping out of Harvard to start a little software company. Microsoft's first product was a Basic compiler for the Altair 8800 crowd. And Steve Jobs and Steve Wozniak were starting Apple computer out of a garage. We could follow what everyone else was doing by reading Byte Magazine. As far as I can tell, I was right in there, playing around with primitive personal computers at the same time Steve Jobs and Bill Gates were starting out.

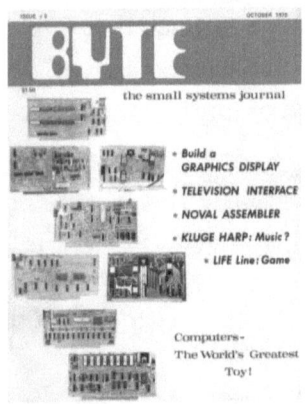

I felt very fortunate that I had a good paying management job at Motorola, so it never really occurred to me to change careers. But I probably should have gone to work for Apple or Microsoft or one of the other early personal computer companies. The pioneers in the industry all went on to become fabulously wealthy!

6) Over the course of the 1½ years that all the above was going on, I gradually added more memory to my computer. The original kit came with a memory card that held 4K bytes of static RAM on 32 chips of 1K bits each. I ordered a second memory card and finally a third one, and by this approach got all the way up to 12K bytes. This was great compared to the original 4K bytes, but even then, lack of memory was a perpetual problem. Something had to be done, but I was also running out of card slots on the motherboard.

I decided to add 32K bytes of additional memory on a single card by building my own dynamic RAM card. Using dynamic RAM, I could buy 16K bit chips. Sixteen of them would give me the planned 32K bytes. However, the board also would need several control chips to handle the refreshing of dynamic RAM and the whole thing needed buffers between the chips themselves and the computer bus. There was no kit for this. I bought a SW Technical prototyping board and designed the whole thing myself. I had to do a lot of studying to figure out how all this stuff worked and then design the circuit and the circuit board. It was a huge relief when I got the whole thing done and it actually worked. I now had 44K bytes of RAM. Although that doesn't sound like much today, it was enough to do a lot of things. To put it in perspective, the first MAC, which didn't come out until 7 or 8 years later, still only had 128K bytes of RAM.

7) The biggest thing still missing from my computer, and the thing that kept it from being a practical tool, was the lack of good mass storage. While I finally had a decent amount of RAM, the only way of storing anything was on a slow, unreliable audio tape system. What I really wanted was a disk storage system. There were three big problems. First, there were no inexpensive hard drives, i.e. less than $10,000. Floppy disk drives were in their infancy. All you could get was a huge 8" floppy drive that held about 400K bytes per disk, and it cost about $500. Secondly, even if I bought one, I had no way of interfacing it with my computer. I would need to build a floppy disk drive controller myself. Thirdly, I had no disk operating system for my computer, so I would have to create one myself.

Fortunately, I knew a guy at Motorola who worked with the Microprocessor System Applications team. He helped me procure an 8" floppy disk drive directly from Motorola for about $400. He also provided me with a schematic for a disk controller and enough information about interfacing it with my computer to get me started. This was an unbelievably large project by today's standards. The disk controller required about 50 ICs all wired together correctly. It also contained a phase-locked loop circuit that needed to be tuned properly to read data from the disk.

I decided to use a "wire-wrap" card to build it on, as it was probably the most reliable way to prototype a complicated circuit. It took me a couple of weeks to round up all the parts and another couple of weeks to actually build the controller. It then took another couple of weeks to get it tested and tune the phase-locked loop. For this last part, I got some help again from my friend who gave me the circuit. To my amazement, since there were 10,000 ways to make a mistake on this very complicated circuit board, the thing actually worked.

This image of a wire wrap card is not
my disk controller, but you get the idea!

8) Now I was ready for the next big project: writing my own disk operating system. It came together slowly over 2 or 3 months. At first, I had to just deal with the basics: formatting the disk, writing data into a single sector (a sector was a block of 2K bytes at that time), and reading back the content of a single sector.

When I had all that working, I did some research on how disks themselves and their operating system are organized. I found that traditionally hard drives sectors were grouped into larger units called clusters, and a "cluster allocation table" controlled how files were allocated clusters and how space on the hard drive was allocated in general. Several other tables were used to make this whole system work. File space was dynamically allocated, and a single file could end up spread out all over the physical disk.

I decided all that was too complicated and too much trouble. I used a single table with the name of the file, the number of sectors allocated to it, and at which sector the file was started. The file was then laid down in sequential sectors until the whole file was stored. When I got this basic scheme implemented, I was finally able for the first time to store a file on my floppy disk and retrieve it. To store a file, let's say my Animals game, you only needed to type: Store Animals 2075 3063, where the 2075 and 3063 were the beginning and ending memory addresses of where the program was currently loaded in RAM. To retrieve the program and start running it, all you needed to do was type: Load Animals. This system had an advantage not only in its simplicity but in speed as well. Programs were stored and loaded virtually instantly, mostly because they were stored on sequential sectors. This may not seem like a big deal to you now, as our computer seem to store files almost instantly, but in the early days of PCs and MACs, storing or retrieving a file could take up to 30 seconds, due to a combination of slow processors and overly complicated disk operation systems.

There were still some other issues that needed to be addressed. A file could be modified and restored in the same place, but if it grew in the process of being modified, it would no longer fit in the same number of sectors as it did before. So the space it previous occupied was marked as unused, and the file was saved in a new location at the end of the used space on the disk. However, this caused the disk to fill up too quickly, leaving behind lots of unused and unusable space. The fix was pretty simple – I wrote a utility program that pulled everything forward, filling all the unused space, and freeing up space at the end that was then available for use again. This utility just needed to be run once in a while when the disk got 70 or 80% full. It worked great.

This was pretty much the final version of my first home computer. The whole project had taken about 2 years, so it must have been completed by early 1978. With the floppy disk drive, 44K bytes of memory, and a simple disk operating system, it was possibly one of the most sophisticated home computer systems in existence, at least for a short while.

By 1978, everyone was jumping on the bandwagon. Hundreds of companies were by now in the process of developing home computers. It would only be another 3 years until the first IBM PC was introduced.

It had been a great adventure!

Some Personal Notes on the "Personal Computer Pioneer??" Story

There are several things of a personal nature that I wanted to talk about as I told my story about the early days of personal computers. They seemed a little off-topic as I tried to include them in the main story, so I am breaking them out here as a separate chapter.

Personal Life

I am going to talk first about how all my efforts with personal computers effected my personal life, or probably more specifically, my wife Donna's personal life. I probably spent 20-25 hours a week on my hobby. That doesn't sound too bad, but that was on top of working 48-50 hours a week.

My typical weekday consisted of getting home from work around 6:00 PM, having a drink with Donna and talking about our respective days. We would then eat dinner, and, after that, I would head for the extra bedroom which I had converted into my computer lab. From 8:30 PM or so until 1:00 AM, I was working on computer stuff. By 6:30 AM the next morning I was up to go to work again, so I was probably not getting enough sleep!

Weekends were a little different, but very hectic. We usually went out with friends on Friday nights and drank quite a bit, so no computer work on Friday. I was a private pilot, so I tried to get down to the airport to fly for an hour on some Saturday mornings. We had a boat and loved to go to the lake, so that happened maybe once every other weekend. Maintaining a boat was a huge time consumer, so a lot of times I was working on the boat on the weekends, as well.

In between all of that, I was working on my computer hobby every spare minute. Donna was pretty frustrated during this time, partly because our life was so busy and so hectic, and partly because I was pretty much an "absentee husband" when I got engrossed in my computer stuff.

Learning to Program Microprocessors

I want to briefly talk about how I learned assembly language programming in order to program the 6800 microprocessor.

Motorola had introduced the 6800 and I knew I wanted to learn how it worked and how to program it. The learning process was very stressful but also interesting! Before I get into all this, however, I need to start with a little background information.

Assembly language is the instruction set hardwired into a microprocessor chip. It consists of instructions like: load the following 8 bits into register A; increment (increase by 1) the content of register B; if the content of register C is greater than some binary number, then branch program execution to the following memory address. Though I was familiar with programming in a high level language, I had no idea how to make anything useful happen using a set of instructions like this.

The 6800 Programming Manual

```
00200 03E8 96 02    MULT    LDA A  NUM1     ;NUM1=MULTIPLIER
00210 03EA D6 03             LDA B  NUM2     ;NUM2=MULTIPLICAND
00220 03EC 7F 0000           CLR    ANS1
00230 03EF 7F 0001           CLR    ANS2
00240 03F2 0C                CLC
00250 03F3 CE 0007           LDX    #7       ;LOOP COUNT
00260 03F6 97 04             STA A  NUM1A
00270 03F8 4F                CLR A
00280 03F9 76 0004           ROR    NUM1A
00290 03FC 5D        YY1     TST B           ;SET COND. CODES ACC.TO B
00300 03FD 2A 02             BPL    YY2      ;CHECK FOR A 1 IN BIT
00310 03FF 9B 04             ADD A  NUM1A    ;OVERFLOW NOT POSSIBLE
00320 0401 0C        YY2     CLC
00330 0402 76 0004           ROR    NUM1A
00340 0405 58                ASL B
00350 0406 09                DEX
00360 0407 26 F3             BNE    YY1      ;CONTINUE UNTIL X=0
00370 0409 97 04             STA A  NUM1A
00380 040B CE 0008           LDX    #8       ;LOOP COUNT
00390 040E 96 02             LDA A  NUM1
00400 0410 97 04             STA A  NUM1A
00410 0412 4F                CLR A
00420 0413 D6 03             LDA B  NUM2
00430 0415 0C        YY3     CLC
00440 0416 56                ROR B
00450 0417 24 07             BCC    YY4      ;IF CARRY,INCREMENT ANS2
00453 0419 9B 04             ADD A  NUM1A
00456 041B 24 03             BCC    YY4
00460 041D 7C 0001           INC    ANS2
00470 0420 78 0004   YY4     ASL    NUM1A
00480 0423 09                DEX
00490 0424 26 EF             BNE    YY3      ;CONTINUE UNTIL X=0
00500 0426 97 00             STA A  ANS1
00510 0428 39                RTS             ;FINISHED, EXIT TO MAIN
00600                        MON
```

Sample of 6800 Assembly Language

I ordered 2 books from Motorola, one on the 6800 itself and its
instruction set, and a second one that was a huge book
documenting how an end product was designed using the
6800, and how all the software was designed to make that
product work. I started trying to read these two books, but it

was as if they were written in a foreign language! The assumption seemed to be that the reader was already familiar with assembly language and all he or she needed to know was the particulars of how this stuff all worked specifically on the 6800.

The learning process during the next few weeks was interesting, because I started reading these two books with almost zero comprehension. I would read for a while, get totally frustrated, then go back and re-read a chapter over again. Very slowly, I started understanding more of the words and terminology, and very slowly, I started to comprehend what I was reading. At some point, after a couple of weeks of struggling, everything started to click; the pieces all started falling into place. I started to understand how all this stuff worked.

This experience was wake up call for me, because up until then, I believed that, if I read something and didn't understand it, I probably never would. Since then, I have had several other encounters with something I found incomprehensible to begin with, only to find out that with perseverance, I would eventually understand it.

Presentation and Newspaper Article

In November 1977, I had an opportunity to give a talk on the current status and future of personal computing. It was at Motorola's Annual Technology Symposium. The symposium was held at Scottsdale Center for the Arts, and was open to Motorola employees, other guests, and a few reporters. There were three of us interested in speaking on the issue, so Personal Computers was designated as one of the major topics.

My talk covered the experience of building and programming a personal computer. I focused on the idea that personal computers were going to be big in the future. I assured my audience that although you could not yet do much with these small computers, that they were already a great adult toy and that it was all worth the effort.

After my talk, a reporter from the Arizona Republic asked to interview me, and came to my house with a photographer as well. The article below was the result. It must have been a slow news day, because it came out on the front page of the Arizona Republic on November 14, 1977:

Ultimate adult toy

Logic of computer hobby is hard to explain

By GRANT E. SMITH

"People keep expecting me to tell them a practical application for it," said Doug Domke.

The shrug of his shoulder and the tone of his voice indicated he regarded the matter as trivial.

After all, Domke had already described his homemade computer as the "ultimate adult toy." Why should a toy be practical?

It was plain to see Domke had developed the same type of relationship with his computer that other men have developed with a hunting dog, an old car, a ham radio set or any other hobby that might strike their fancy.

Domke proudly told about working out games to play with the computer, but added his wife and friends did not think much of the games.

Domke didn't seem to care what others thought. What mattered was that he could communicate with the computer and it could do things. What it could do wasn't important as long as it could do things.

Domke is one of a developing breed of computer hobbyists. Right now, the breed is mainly limited to engineering and technical types, but there are signs the hobby could break loose and infect the nation.

Some estimate the computer hobby may be a $200 million-a-year market by 1982.

Domke, a Motorola engineer, and three of his colleagues described the craze last week at a gathering of engineers from Motorola's Semiconductor Group.

Mike Kabealo pointed out that there are currently about 200 companies making computers for the home.

"They started as kits that you had to put together," he said. "But now, they're sold as complete units."

More than 700 stores across the U.S. are selling them, he added, and 15 publications are geared toward the hobbyists.

"The stores are like the old general store — a place where you can hang out and talk about computers." he said.

Currently, more than half the market

Continued on Page A-10

Republic photo by Nyle Leatham

Motorola engineer Doug Domke of Scottsdale demonstrates use of his homemade computer.

Computer hobby hard to explain

Continued from Page A-1

is in calculators, But Kabealo predicted that by 1983 the market would be dominated by computers that play games.

The ability to add, subtract and play tennis on a television screen has opened the doors of many American homes for the computer.

As technology becomes less complicated and expensive and home computer capabilities increase, computers are expected to gain wider acceptance.

Ed Tynan predicted future families may plunk down as big a chunk of their income for a computer as they now do for a car.

And, the variety of things the computers may do could be endless.

They could balance the checkbook, water the lawn and garden, prepare the family menu, inventory the contents of the freezer, prepare the income tax forms, maintain the proper chemical balance in the swimming pool and control heating and cooling.

Tynan even described a computer-run alarm system that could broadcast a tape-recorded message to the police when the house is being burglarized.

But those are practical applications for a much more domesticated computer than the alley cat now being played with.

Just as a past generation sat patiently through the night wearing ear phones to hear a few garbled words spoken halfway across the continent, the computer hobbyists communicates with his homemade computer through an intricate code of electronic impulses.

In the old days of radio, the actual message often meant less than the fact that it was received at some distant point. The same is true for the hobbyist and his computer; what counts is that the computer responds to the message.

A satisfactory response can take many forms.

Before he got a video display unit for his computer, Domke remarked, "You had to content yourself just to watch the lights on the front panel blink."

If You're Interested:
A Computer's CPU

At the heart of every computer is a CPU or "central processing unit". It's the guts or inner workings of the computer that actually allow it to do what it does.

Programmers learn to write programs that are executed by the CPU, but what actually happens in the central processor is a mystery even to most programmers, unless they are specifically writing "assembly language" code.

My own personal experience with CPUs began when trying to understand the inner workings of Motorola's 6800 micro-processor back in the mid 1970's. A microprocessor is basically a CPU built on a single integrated circuit chip.

The photo above is what the 6800 looked like from the outside. The photo below shows the chip on the inside. It is a complicated circuit made up from 6800 transistor elements.

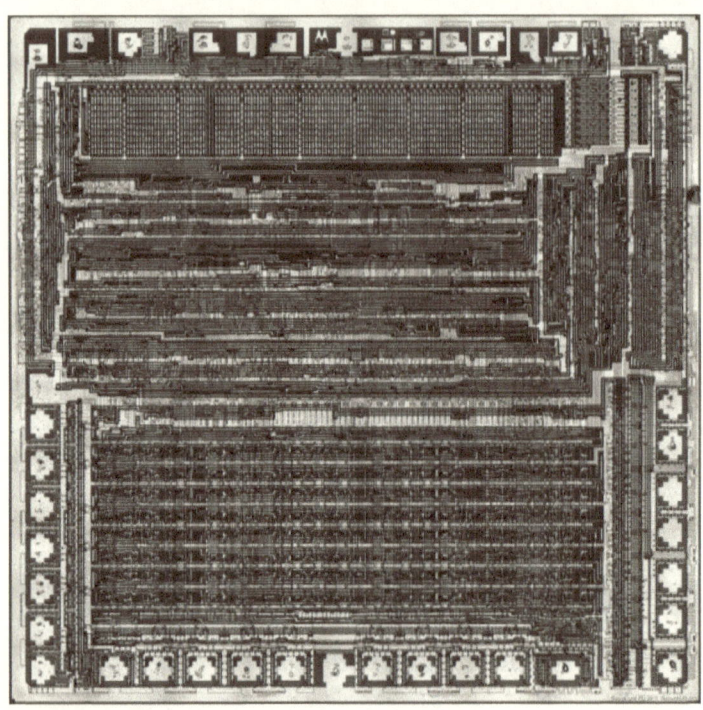

The 6800 was leading-edge technology back in 1974, and putting 6800 transistor elements on a single chip was a major accomplishment back then. However, the modern processors that power today's PCs, MACs, and cellphones are far more powerful than the 6800. They are also much, much more complicated, containing as many as 100 million transistor elements!

But all CPUs basically operate the same, and the stuff they do is really pretty simple. They do it really fast, typically fractions of a microsecond. That means they can perform tens of thousands of operations in a thousandth of a second. But what they do in any one operation is not complicated. And CPUs are also simple in the sense that everything they do is done sequentially, one operation at a time. Toward the end of this chapter, after we have defined some of the pieces that make up a CPU, we'll come back to what all those 6800 transistor elements actually do.

Let's begin with the concept of a register. It's a place to store one number. Numbers in CPUs are binary, meaning they are composed of 1's and 0's. A single 1 or 0 is known as a bit, and 8 bits is known as a byte. So, for example, a 16 bit register can store sixteen 1's or 0's, and those sixteen 1's or 0's represent a number between 0 and 65,535 (2^{16} possibilities).

CPUs have several different registers. One called the program counter register points to the instruction in memory that is currently being executed. We will talk about program instructions and their execution in a moment. Some other registers called accumulators store numbers for arithmetic processing, such as addition or multiplication. They can also just be a place to temporarily store a number. Other registers such as index registers and stack pointers are places to keep track of something, like where we are in a list that is being processed.

Let's now deal with another basic concept – the bus. A bus is a physical pathway, usually between a register in the CPU and a location in memory. It is a certain number of bits wide, meaning it handles that many bits at once. It could be 8 bits wide and handle numbers between 0 and 255 (2^8 possibilities). Or it could be 64 bits wide and handle numbers between 0 and 18,446,744,073,709,551,615(2^{64} possibilities).

One of these buses is an address bus; it's used to connect the CPU to a specific location in memory. Another bus is a data bus; it is used by the CPU to read data from or write data into a specific address in memory.

Now let's talk about the processor's instruction set. It's a list of operations the processor is capable of carrying out. These operations are all simple stuff. The power comes not from any one instruction, but from the ways they can be ingeniously combined to accomplish complex tasks. For example, one instruction might say "load accumulator B with the data stored at address 23454321". Another instruction might be "increase the content of accumulator B by 5". And another could be "put the content of accumulator B into the data stored at the memory address currently pointed to by the index register".

To see how you might get something useful done with instructions like this, let's suppose we want to output a line of text to a printer. The text is stored in memory starting at address 345678 and is 28 characters long. A character might be one letter, a period, or a space. Each character is stored in one byte of memory (8 bits). Each byte has its own address in memory, so if the first character was stored at address 345678, then the second character would be stored at 345679, etc. Let's also assume that there is an 8 bit register called print-out. Data sent to the print-out register is automatically passed on to the printer. Let's see the kind of CPU instructions that would cause our text to print.

1. Load index register with starting address of the text: 345678.

2. Load accumulator A with the number of characters: 28.

3. Load accumulator B with content pointed to by the index register: the first character in our text

4. Store the content of accumulator B in the print-out register.

5. Increment index register by 1.

6. Decrement accumulator A by 1.

7. If accumulator A content is greater than 0, go back to step 3 and repeat.

This code would loop around 28 times – each time reading one character and sending it to the printer. The CPU would actually have to execute 142 operations in order to print our text.

The programming language of the CPU is usually called assembly language. A program called an assembler is used to write assembly language. An example of assembler code for the 6800 microprocessor is below. (It's just a sample to see what assembly language looks like – not our code to print a line of text.)

```
C000                         ORG     ROM+$0000 BEGIN MONITOR
C000 8E 00 70   START   LDS     #STACK
C003 86 13      INITA   LDA A   #RESETA   RESET ACIA
C005 B7 80 04           STA A   ACIA
C008 86 11              LDA A   #CTLREG   SET 8 BITS AND 2 STOP
C00A B7 80 04           STA A   ACIA
C00D 7E C0 F1           JMP     SIGNON    GO TO START OF MONITOR
C010 B6 80 04   INCH    LDA A   ACIA      GET STATUS
C013 47                 ASR A             SHIFT RDRF FLAG INTO CARRY
C014 24 FA              BCC     INCH      RECIEVE NOT READY
C016 B6 80 05           LDA A   ACIA+1    GET CHAR
C019 84 7F              AND A   #$7F      MASK PARITY
C01B 7E C0 79           JMP     OUTCH     ECHO & RTS
C01E 8D F0      INHEX   BSR     INCH      GET A CHAR
C020 81 30              CMP A   #'0       ZERO
C022 2B 11              BMI     HEXERR    NOT HEX
C024 81 39              CMP A   #'9       NINE
C026 2F 0A              BLE     HEXRTS    GOOD HEX
C028 81 41              CMP A   #'A
C02A 2B 09              BMI     HEXERR    NOT HEX
C02C 81 46              CMP A   #'F
C02E 2E 05              BGT     HEXERR
C030 80 07              SUB A   #7        FIX A-F
C032 84 0F      HEXRTS  AND A   #$0F      CONVERT ASCII TO DIGIT
C034 39                 RTS

C035 7E C0 AF   HEXERR  JMP     CTRL      RETURN TO CONTROL LOOP
```

Programmers today typically write software with high-level language compilers like C++, Python, or Java. They are powerful tools that make it much easier to program complex tasks. The compiler processes the high-level language code and translates it into machine code. The programmer doesn't need to know assembly language or what the CPU is actually doing. In the high-level language, he would just write the code:

Print "This is a sentence to print."

The compiler would then generate the machine code required. (Note our sentence above has the same 28 characters we used in our earlier example. Spaces and the period are themselves characters.)

Our assembler or our high-level language ultimately creates machine code – the stuff that is actually executed by the CPU. What does machine code look like? An example is below.

```
3F20   3F 20 3F F6 0C 8E A0 60 CE 3F B7 A6 00 C6 2E 11
3F30   27 06 BD E1 D1 08 20 F3 86 0D BD E1 D1 86 0A BD
3F40   E1 D1 86 00 BD E1 D1 BD E1 D1 CE 3F CF A6 00 11
3F50   27 06 BD E1 D1 08 20 F5 8D 51 FF 3F 20 BD E0 CC
3F60   CE 3F E3 A6 00 C6 2E 11 27 06 BD E1 D1 08 20 F3
3F70   8D 39 08 FF 3F 22 FE 3F 20 86 0D BD E1 D1 86 0A
3F80   BD E1 D1 86 11 B7 3F 24 FF 3F 20 CE 3F 20 BD E0
3F90   C8 FE 3F 20 7A 3F 24 27 E0 BD E0 CC A6 00 BD E0
3FA0   BF BC 3F 22 27 02 20 EC 7E E0 E3 BD E0 CC 86 3F
3FB0   BD E1 D1 BD E0 47 39 48 45 58 41 44 45 43 49 4D
3FC0   41 4C 20 4D 45 4D 4F 52 59 20 44 55 4D 50 2E 46
3FD0   49 52 53 54 20 42 59 54 45 20 54 4F 20 50 52 49
3FE0   4E 54 2E 4C 41 53 54 20 42 59 54 45 20 54 4F 20
3FF0   50 52 49 4E 54 2E
```

Machine code is basically a list of numbers. They are shown in the example above in hexadecimal (base 16) format. (Why hexadecimal? It allows us to express an 8 bit or 1 byte number as two digits - you can't do that with decimal, as there are 256 possible numbers in a byte.) Some of the numbers are actual instructions, while others are data to load, memory addresses, etc. As the program counter goes through this list, the processor is able to decode it into actual instructions and execute them. For example, 86 is an instruction for the 6800 that says "load accumulator A with the byte that immediately follows the instruction itself". If the next byte is 3F, which is hexadecimal for 63, then 63 would be loaded into accumulator A by this instruction. In general, machine code is specific to a particular CPU, so it only works with the specific CPU it was written for.

Now the next thing you might want to ask is "How does the CPU actually do any of this?" Everything is basically hard-wired into the chip. The registers, the busses, the logic required to execute the instruction set, the arithmetic processes, etc. are all hard-wired circuitry built into the integrated circuit. Those 6800 transistor elements are wired together to make the registers and other physical hardware needed to make the CPU work.

So now you know a little bit about the CPU at the heart of a computer.

Commercial PCs
1975 -1980

A few chapters ago, we talked about commercial computers beginning to appear in 1975 when MITS introduced the Altair 8800. It was a kit you had to build, and was really only intended for computer hobbyists. A whole bunch of other kits and even assembled computers came out over the next three years, but they didn't really do much and were intended for hobbyists, rather than ordinary consumers.

The focal point of the industry during the first three years after the Altair 8800 seemed to be the Homebrew Computer Club in Silicon Valley. It's where Steve Jobs and Steve Wozniak showed off their Apple I computer. Like Apple, several other early computer entries could trace their origins back to the Homebrew Club.

And one other significant development occurred during this period. Bill Gates and Paul Allen wrote a BASIC compiler for the Altair 8800, and proceeded to found Microsoft Corporation.

Commercial computers finally started to change into something people could actually use at the end of 1977, with the introduction of three new systems: the Apple II, Commodore's PET, and Radio Shack's TRS-80. We will look briefly at each of these.

The Apple II was perhaps the first complete computer. It had color graphics, a full QWERTY keyboard, and internal slots for expansion, which were mounted in a high quality streamlined plastic case. The monitor and I/O devices were sold separately. The original Apple II operating system was only a built-in BASIC interpreter contained in ROM. The Apple II used MOS Technologies' 6502 processor and started out with only 4K bytes of RAM memory.

Sales started out slowly for the Apple II. Its price was higher than the other 2 systems, and its BASIC lacked floating point math. Apple lacked any retail distribution system at that point which added to their problems.

But the Apple II was eventually a huge success. In the next couple of years, floppy disk systems for the Apple II became available, and the operating system got upgraded to a disk operating system. Amazingly, the Apple II remained in production until 1993, and 4 million systems eventually shipped.

The Commodore PET (short for Personal Electronic Transactor) also used MOS Technologies' 6502 processor, just like the Apple II. It was a single-board computer with built in video hardware driving a small built-in monochrome monitor with 40 × 25 character graphics. The processor card, keyboard, monitor and a data-cassette for data storage were all mounted in a single metal case.

The PET shipped initially in two versions, one with 4 kB of RAM, and another with 8 kB. The built-in data-cassette was located on the front of the case, which left little room for the keyboard. The machine was fairly successful, but there were frequent complaints about the tiny calculator-like keyboard.

Although the keyboard issue was eventually fixed, and there were several other upgrades over the next couple of years, the PET was the least successful of the three 1977 entries with total sales never exceeding 1 million units.

Radio Shack introduced the TRS-80 in late 1977. As improved models were introduced later, this first version became known as the Model I. The Model I combined the motherboard and keyboard into one unit with a separate monitor and power supply. The PET and the Apple II offered more advanced features than the TRS-80, but Radio Shack's 3000+ stores gave it a distribution network that neither Apple nor Commodore could touch.

The Model I used a Zilog Z80 processor and shipped with 4 kB of RAM. It had a full size QWERTY keyboard, well written Microsoft floating-point BASIC, and the inclusion of a monitor and tape deck for approximately half the cost of the Apple II. The Model I added an optional floppy disk drive a couple of years later. The Model I was discontinued in 1981, having sold 1.5 million units. Radio Shack did, however, introduce a Model II and Model III, so the TRS-80 stayed around for many more years.

So by 1978 we had a few personal computers that started to sell to the general public. However, there were three rather obvious limitations to personal computer progress. Two of them I already discussed at length in the chapter about my own struggles building a usable computer.

Notice that all these new arrivals started out with only 4K bytes of RAM. That is a serious limitation to doing anything practical. Adding more memory required a lot of extra boards and cost. And they all started out with another problem – lack of good mass storage. The digital cassette approach was better than nothing, but left a lot to be desired. Finally starting to be available as add-ons in the late 1970's, 5 ¼ inch floppy disk drives eventually solved this issue.

A third problem that we haven't talked about yet was software. Hobbyists like me were fine writing their own programs, but the general public wanted applications software that allowed them to put these personal computers to practical use. They finally started to appear in the late 70's. Two in particular stand out as significant.

In 1978, Micropro International introduced WordStar, the first successful word processor. It originally was written only for CP/M, the operating system of the Intel 8080 based machines. But later it became available for the TRS-80 and other systems, eventually becoming an industry standard that was only gradually replaced by Microsoft Word.

In 1979, Visicorp introduced VisiCalc, the first spreadsheet program. Initially, it only ran on the Apple II, but was gradually made available on other operating systems. VisiCalc remained popular for several years, eventually being replaced by Lotus123 and then Microsoft Excel.

By 1979, personal computers typically came with more than 4K bytes of RAM memory - maybe 12K, 16K or even 28K. But even so, VisiCalc made these early computers' RAM limitations blatantly obvious. Most systems would have crashed or generated a "Memory Full" error if you attempted to generate a 50 x 50 cell spreadsheet!

While we are on the subject of software, we should talk a little more about operating systems. They were pretty primitive in the beginning with one exception. Even in 1975, MITS and IMSAI 8080 computers could use a disk operating system called CP/M available from a company called Digital Research. It took a few years before floppy disk drives were affordable for hobbyists, but CP/M was a full blown disk operating system that was available when 5 ¼ inch floppy disk drives came along. While other disk operating systems appeared over time, they were generally modeled after CP/M, including MS-DOS.

Computers continued to improve and new models kept being introduced, but the next big development was the introduction of the IBM PC and Microsoft's DOS operating system. That will be the subject of an upcoming chapter.

Progressive Business Computers

Another interesting computer adventure occurred for me around 1980, when I joined a friend of mine in starting a small company selling business computers. This tale is slightly off subject, but it does give you a glimpse into the state of the industry in 1980.

My friend was a co-worker at Motorola named Jim Griffin. He had by this time left Motorola, and was trying to make it on his own as a small business entrepreneur. He had a small business designing and installing private phone systems for small companies, schools districts, etc. Jim was an extremely smart guy. While I worked with him at Motorola, he became fascinated with a computer language called APL. I thought it was the most bewildering computer language I had ever seen. It was almost impossible to decipher a finished program, causing people to jokingly call it a "write only" language. Jim became an expert in this language. Later, Jim ended up managing the entire Information Systems Department of Motorola Semiconductor. His background was semiconductors, not computers, but he was able to head-up the Information Systems organization in spite of that!

Jim wanted to start up a computer company that would supply computer hardware and software to small businesses. He needed a partner with a good knowledge of small computers. He and I founded Progressive Business Computers. I had a 49% ownership and was to continue working at Motorola, with no need to make any money from this business anytime soon. Jim had a 51% ownership and needed the business to start generating some income for him in the near future. Had either one of us ever been in a business partnership previously, we would have recognized this as a recipe for disaster before we ever started.

While I don't remember the exact year we started this, it was when computers such as Radio Shack's TRS80 and the Apple II were around, but well before the IBM PC was available. At this time, business computers and business software were the domain of big companies like IBM. If you wanted a computer to run business systems on, you bought a $300,000 IBM mainframe and then $500,000 worth of software and peripherals to make it all work. There really weren't any low cost, affordable business computers out there. A small business had a bookkeeper or accountant, who kept the books using pencil and paper along with a calculator – spreadsheets had recently appeared, but weren't in widespread use yet.

A few companies were developing small computers suitable for a small business, and others were working on software packages for accounting, etc., but at this point, there were no established products out there. No one had put it all together!

Our first task was to find a computer we could base the business around. We finally agreed upon a very robust and professional looking computer built by Tano Corporation. They were selling just what we were looking for, a computer designed for business, rather than the hobbyist crowd.

I couldn't find a photo of the Tano,
but it looked something like this!

The Tano was encased in a very rugged steel enclosure. It looked more like a tank than a computer. It was so heavy, you could barely lift it. It came with a 5¼" floppy disk drive and a Basic interpreter, so that it was possible to write programs in Basic, and store data and applications on floppy disks. I don't remember details of the processor, amount of memory, etc., but it was adequate to actually run small business applications. It was much more expensive than a hobbyist computer, so even buying a single development system was a major investment for us. We also needed a printer for the minimum business system. The only affordable printers at that time were dot-matrix printers. We bought a fairly rugged model, big and very noisy by today's standards.

Now that we had agreed on the hardware, we had to find some business software. We hardly knew where to start. But there were a lot of small start-up companies trying to figure out the same things we were, and after some research, we found a company in Utah, Great Plains Software, that had a suite of business and accounting programs, all written in Basic. However the Basic computer language was never standardized, back then or even today. There were hundreds of variations, and the language tended to take on a new variation with each set of hardware that it ran on. So getting Great Plains software to run on the Tano Computer looked like a major project. Jim went to Utah, met with the people at Great Plains, and came back with an agreement that they would give us the software in exchange for our creating a version of their software that could run on the Tano.

This is where I got very busy. It was my task to rewrite all the software, and get it working on the Tano system. Jim's job was to find potential customers. The task of converting this huge set of applications to allow them to run on the Tano turned out to be about 10 times as hard and time consuming as I had originally thought. I began the task thinking it would take 2 or 3 weeks of evenings to get it done. It was more like 6 months of evenings and weekends.

Jim found a couple of potential customers, and was stringing them along, while I was trying to get the applications all converted. I think he managed to get a couple of small consulting jobs along the way which gave him some income while all this was going on, but we were both getting extremely frustrated with this whole arrangement. Jim was frustrated because he didn't have enough income to provide for his family. I was frustrated because I had put 500+ hours of work into this conversion and had no prospect in the foreseeable future of ever seeing any return from it.

I finally finished the conversion, and we finally had something saleable. It was roughly like a personal computer with QuickBooks installed on it, but it was before we had QuickBooks or affordable personal computers. Not too bad for the late 1970s!

However, we still didn't have a customer that was actually ready to buy a system. Jim by this time was much more interested in doing anything for anybody that brought in some short-term income – I don't think he was very focused on selling our systems. Several more weeks went by, and still no sales. We both grew more frustrated, both with the situation and each other. Jim wanted me to put more money into the business to keep things going and to let him take some money out. I didn't want to do that; I was starting to think this whole thing was doomed. Finally, Jim could not wait any longer and decided to get a job. We both pretty much gave up on making this work!

There were some lessons here, I guess. We were both smart guys, even pretty smart about business, and we put everything regarding our partnership in writing. But there were a couple of areas we didn't spend enough time thinking through.

First, the market wasn't quite ready. Both the hardware and software were only barely capable of meeting the demands of business at that time. Small businesses were just not ready to shell out thousands of dollars on a computer system. And medium sized businesses had by then found big accounting firms with mainframe computers that could handle their books.

Perhaps even more to the point, we didn't really comprehend the effects that our different objectives would have on how this all played out. We didn't put enough money into the business – Jim didn't, because he didn't have it at the time, and I didn't, because I was not ready to risk a lot of money on something that might or might not work out. And we certainly didn't allow enough time or put enough effort into our business to make it successful.

We dissolved our business without ever making a sale. I got to keep the Tano in exchange for all my labor, though it was small consolation for that much effort. The whole thing was a bitter experience for both of us, but even more so for Jim, I think. At least I had kept my good paying job at Motorola through it all.

Jim and I stayed in touch for a few more years, but we were never again the good friends we had been before starting this business.

The IBM PC and MS-DOS

In the mid-70's, the computer giant IBM simply could not figure out what to make of the new micro-computer industry. It appears they viewed it almost like a joke to start with, but gradually began to view it more and more as a serious threat to their business. But did they really want to get into the consumer electronics business? They didn't know!

Finally in 1980, IBM decided to enter the market. From the perspective of the personal computer enthusiast and potential business customers, IBM's entry into the market was fantastic. It gave the entire new industry legitimacy! It also meant a lot more potential software development, as IBM's main goal was to sell personal computers to businesses.

To its competitors, however, IBM looked like the "Evil Empire". IBM typically dominated any market they decided to get into, and put a lot of smaller companies out of business. As it turned out, however, the IBM PC only partly lived up to the hopes of its customers and fears of its competitors. Let's first look at the machine that came out in September of 1981 – the Model 5150.

The base price was $1565, and it came with 16K bytes of RAM and audio cassette for data storage. No monitor was included. It used the Intel 8088 processor, which had16 bit registers and internal data paths, but only an 8-bit external data bus.   It came with Microsoft Basic, but no other software or operating system. In this minimum configuration, its performance was typical of what had been around for the past four years. However, there were options that brought it up to speed with the latest entries. You could buy it with 64K bytes of RAM, a floppy disk drive, a monitor, and Microsoft PC-DOS for about $3000. Add a second floppy, color graphics, and a printer for about $4500.

So the IBM PC wasn't cheap. In today dollars (2019), the price would range from $5000 - $15,000!

IBM made some interesting decisions in bringing its first personal computer to the market in 1981. Those decisions had a huge impact on the direction of the industry. Much of that impact is still with us today. IBM typically used highly proprietary technology for its mainframe products, so that its hardware was covered by a myriad of patents, and its software was all treated as secret intellectual property.

But with the PC, none of that was true. They were anxious to get something into the market quickly, so they used all standard "off the shelf" components for the hardware, starting with the Intel 8088 microprocessor.

For software, they wanted to license someone else's software – someone like Microsoft who was already in the personal computer business and who understood the market. IBM didn't feel that they understood the business. Their business was mainframes, not personal computers.

The decisions made by IBM had a profound impact on the PC industry, as we shall see shortly. IBM did very well in the personal computer business over the next two decades, but might have done much, much better were it not for these decisions.

Let's talk first about the hardware. As soon as IBM introduced the PC, there were shortages. IBM couldn't make them fast enough. And because they were built completely from off the shelf components, it was extremely easy for other companies to copy the design. Compaq Computer introduced a fully compatible clone of the IBM PC only a year after the PC was introduced. Many others followed. Some like Compaq tried to be 100% compatible with the PC, but others didn't care about exact compatibility. They were content to just use the same Intel processor and Microsoft MS-DOS. Columbia and Franklin were a couple of these early clone suppliers. Others who came a little later like Dell and HP are still familiar names today. The IBM PC market in total, including the clones, quickly deteriorated to a version of the "wild west". There were no rules, and everyone was doing their own thing. The lack of standardization in hardware that resulted is still a problem with PCs even today.

On the software side, IBMs decisions were even more interesting. In fact, whole books have been written about how Bill Gates and Microsoft succeeded in getting and keeping the operating system for the IBM PC. I will try to give you a short version of this strange tale here.

As mentioned already, IBM decided to license someone else's existing disk operating system for their new personal computer. They approached Bill Gates in Seattle, who was well known for his Microsoft Basic. Bill explained to IBM that he didn't have a disk operating system, but that the people they should be talking to were Digital Research, the company which sold CP/M. CP/M was then by far the dominant disk operating system for Intel based personal computers. It had been around for several years, even before affordable floppy disk drives were available.

IBM next attempted to meet with Gary Kildall, the founder of Digital Research and author of CP/M. Gary didn't like or trust IBM, so he left the entire meeting and licensing agreement to his wife and business partner Dorothy McEwen. The first thing IBM wanted was for McEwen to sign a standard non-disclosure agreement, but McEwen, who apparently shared her husband's distrust for IBM, refused. IBM ended up leaving that day without any agreement.

Bill Gates, in the meantime, was busy looking for options, in case he got a second chance at licensing the IBM disk operating system. He found a company right there in Seattle that had a system similar to CP/M. It was called QDOS for "Quick and Dirty Operating System". Gates went to the company, Seattle Computer Products, and succeeded in buying the rights to QDOS for $75,000, without disclosing that he already had IBM as a potential customer.

Bill Gates   Gary Kildall   Tim Peterson

The second time IBM contacted Microsoft, Bill Gates was ready. He ended up hiring QDOS's creator, a programmer named Tim Peterson, and after some minor changes, successfully licensed his new MS-DOS to IBM for the PC.

There are still two more intriguing pieces to the MS-DOS story. The first is that IBM somehow generously allowed Microsoft to retain ownership and control of the operating system. They were content to let PC-DOS, as they called MS-DOS, simply be a Microsoft product you could buy for your IBM PC, rather than insist on buying it outright, as Gates had done with Seattle Computer Products. This turned out to be the single greatest piece of good luck in Bill Gates' life - within just a few years, Microsoft was worth much more than IBM!

The last piece of the MS-DOS story returns to its source: QDOS and its author Tim Peterson. Digital Research and numerous others have made claims over the years that MS-DOS was simply an illegal copy of CP/M, only thinly disguised in an attempt to hide the CP/M code it contained. Tim Peterson readily admitted to cloning the CP/M operating system, but he was quite adamant about the fact that he wrote all the code himself.

Peterson wanted his QDOS to behave like CP/M, and he deliberately matched his APIs or application programming interfaces item for item with CP/M's. But that doesn't imply he stole CP/M's code.

So the battle raged for years with claims that Bill Gates somehow stole CP/M from Digital Research. Some people still believe this to be the case even today. But in all probability, Bill Gates was just shrewd and very lucky. Neither Bill Gates nor Tim Peterson has ever been prosecuted to stealing CP/M.

So IBM's entry into the personal computer market gave the industry a big boost and changed everything. It made the IBM PC and IBM compatible clone PCs the industry standard. By the way, the term PC was around before IBM's entry and just meant personal computer. But once IBM entered the market, PC came to mean just the IBM PC or IBM compatible desktops.

IBM had a very successful position in the personal computer market throughout the 1980s. It introduced a number of new models:

- IBM Personal Computer XT (1983)
- IBM Portable Personal Computer (1984)
- IBM PC Junior (1984)

- IBM Personal Computer AT (1984)
- IBM Personal System/2 Series (1987)
- IBM PS/1 Series (1990)

For businesses, IBM seemed like the obvious choice. By the mid-80s, IBM PCs were in 56% of all businesses that used personal computers. Apple was a distant second at 16%. Together with its clones, IBM and IBM compatibles grew rapidly throughout the 80s, growing to an 85% market share of the entire industry by 1988!

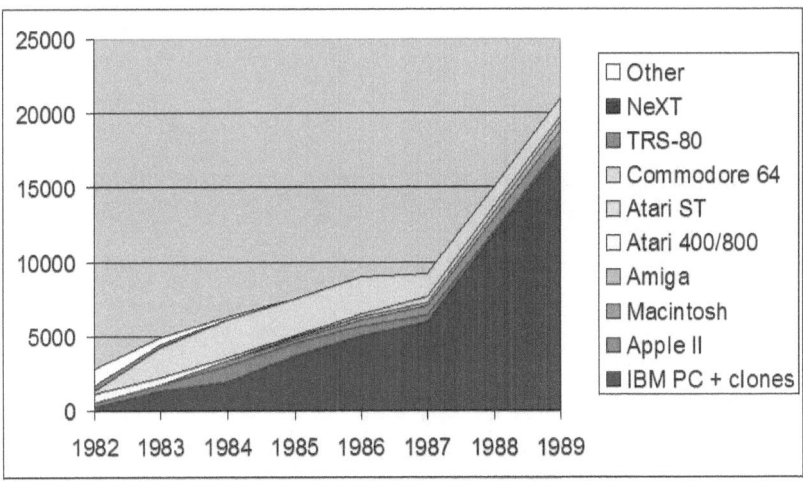

But with all of its success, IBM was not the company that benefited the most from the dominance of IBM computers and their compatible clones. Intel processors became the dominant processors for the personal computer industry and over the next several years made Intel the biggest and most successful semiconductor company. And Microsoft, both with its MS-

DOS and all the supporting applications, like Word and Multiplan (a precursor to Excel), became the dominant software supplier for the personal computer industry.

Intel and Microsoft turned out to be the big, enduring winners. They were not only the financial winners. They controlled the direction of the industry over the next several years. IBM had very little influence in comparison. IBM had a great run – all the way through the end of the 1990s. But as we will talk about later, IBM isn't even in the PC business today.

If You're Interested: Computer Memory

Since we all work with computers today, we can't help but learn a lot of computer terminology. We've all heard about RAM and ROM memory. We know we have a Flash memory card in our cell phone or digital camera. We know there's a hard drive in our computer. We watch movies on DVDs. These are all references to computer memory. But why are there so many different kinds? And what is the purpose of computer memory in the first place? In this chapter, we will attempt to answer those questions, and demystify computer memory.

Let's start with the basic question. Why do computers need memory at all? And what is computer memory?

Computers basically take input data, such as numbers or text, process them, and then output new data. To do this, a computer needs a place to store the data, at least long enough to operate on it, and it needs a place to store the list of instructions about what it is supposed to do with the data. It needs a place to put the result, at least until it's printed, and perhaps to save it for years.

Computers do everything in binary. The data and the instructions are all encoded in long lists of 1's and 0's, so computer memory is simply a place to store 1's and 0's – usually very large quantities of 1's and 0's are needed to do anything useful.

In the earliest days of computers, memory was accomplished by running wires through little circular magnets. By passing current through the wires, these magnets could be magnetized in one direction or the other to store a 1 or a 0. Another wire could then be used to retrieve the 1 or 0 stored in that magnet. One magnet was required for each location where a 1 or a 0 was stored, so to get enough memory to do something useful required filling large rooms with thousands of boards full of wire and magnets.

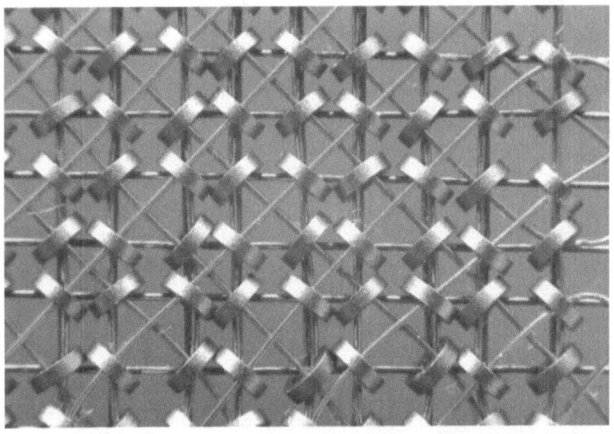

However, aside from being very expensive to build and taking up huge amounts of space, this old "magnetic core" memory had some pretty good properties. It could read or write a lot of 1's and 0's very quickly. It could be accessed randomly, meaning the computer could address any particular location in this vast array of magnets, and read or write a 1 or 0 there. Finally, this memory was what we call "non-volatile", that is, when the computer was shut off and power was removed, the magnets stayed magnetized and retain the stored 1s and 0s.

As we will soon discuss, these are all desirable properties for computer memory.

A big breakthrough occurred in 1970, when Intel introduced the 1103 DRAM chip with 1024 bits of storage on a small, single chip. It replaced a large board with 1000 magnets on it! DRAM stands for dynamic random access memory.

DRAM is still used as the main memory in computers today. That's because DRAM is very dense, meaning we can get the most memory on a chip using DRAM technology. It only requires a tiny, single transistor for each bit stored.

For comparison with the Intel 1103 chip from 1970, today in 2019, we can get 2 GBytes or 16 GBits on a single chip! That's 16,000,000,000 1's or 0's, or 16 million times the capacity of the 1103!

Semiconductor DRAM is pretty fast, and as the name implies, it's random access. But it is volatile; in fact, it's incredible volatile – left to itself, it only can hold its information for a few thousandths of a second, so it has to be constantly refreshed. Refresh means read the data stored before it disappears and then rewrite it in that same location. This constant refresh process requires some special circuitry and means the memory is unavailable to the computer during the refresh process.

DRAM is used for the large working memory of a computer. It's only temporary storage, but it's random access unlike a hard drive. Think about a computer processing an Excel spreadsheet. When you change one number, the computer is able to go around and update all the other numbers on the spreadsheet as a result of your change. This requires being able to quickly move around in memory and implement changes. That's what random access is all about.

The constant need to refresh DRAM slows it down. Because of the refresh process, DRAM is not always available when the central processor needs data or instructions. So another type of memory is used by the central processor for data that is changed quickly and often – it's called static random access memory. It's more expensive and takes up more space on a chip than DRAM, but it is much faster and doesn't need to be refreshed. You will sometimes hear it referred to as the processor's cache memory. Some static RAM is built into the processor itself and some can be externally made available to the processor.

Static RAM doesn't need to be refreshed, but it is also volatile. When power is turned off, static RAM loses its stored information. So next we need to look at the various ways to get non-volatile memory.

The most obvious non-volatile memory that we are all familiar with is the computer's hard drive. They consist of one or more magnetic disks rotating at about 120 revolutions per second.

Magnetic heads can record strings of 1's and 0's in tracks on the magnetic disk. These devices can store incredible amounts of data. A 1TB (one terabyte) hard drive can store 8,000,000,000,000 1's or 0's!

Hard drives are non-volatile. The data on them lasts for many years. They are very reliable, and the least expensive way to store large amounts of data. But they have two major

shortcomings as computer memory goes. They are not random access - data has to be read or written serially in long blocks, so they are not suitable for main memory. And they're also very slow to access compared to DRAM or static RAM.

Another non-volatile memory is CD-ROMs and DVDs. These are similar to hard-drives, but are slower and less expensive. They are optical devices using a laser to reflect light off small pits embedded in the plastic.

Hard drives and CD-ROM/DVD players have one other shortcoming. They are full of mechanical moving parts which wear out and eventually fail. So many other types of non-volatile memory have been developed. These other types are all semiconductor based with no moving parts. There are a number of technologies, and we won't talk about each one, but we will talk about two of the most common.

ROM stands for read-only memory. It's non-volatile, random access, fast, and inexpensive. It's used anywhere that requires a fixed set of computer instructions that aren't going to change. ROMs are hard-wire programmed with those instructions. Smart appliances typically have their program stored in a ROM on the microcontroller. Your personal computer has a small start-up program called the BIOS that starts the process of loading the operating system. It's stored on a ROM.

Flash memory is another type of non-volatile memory. It's fast and inexpensive, can be written and re-written unlike ROMs, but it typically has to be accessed serially, like a hard drive. It is used anywhere that a hard drive might be used. Solid-state hard drives are an example, where a large array of flash memory is used to actually replace the mechanical hard drive in a computer. The memory card in your digital camera, camcorder, and cellphone are also typical applications for flash memory.

So now we have talked about various types of computer memory. We've looked at volatile versus non-volatile. We've looked at random access versus serial access. We've looked at speed versus cost. We've looked at dynamic versus static. Let's try to put it all together and summarize.

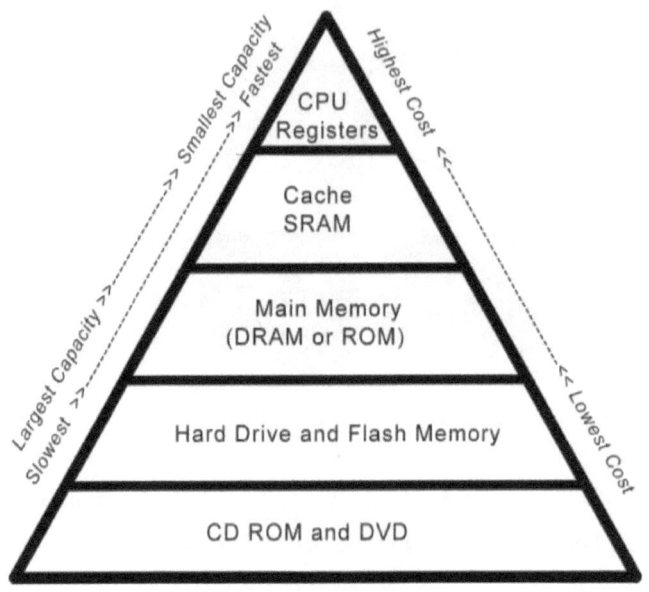

We have many different kinds of computer memory, because the competing technologies are full of trade-offs. We need small amounts of very fast, but expensive memory available to the CPU for speed. We need slower, less expensive memory in large quantities to store our stuff, like pictures, movies, and other data.

So now you know a little bit about all the different kinds of computer memory.

The Apple Macintosh

Following the introduction of the IBM PC in 1981, the next big thing to happen in personal computers was Apple's introduction of the Macintosh in 1984. There is some history which led up to the Mac, and we will get to that in a moment, but the reason the Mac was such a major event in the history of digital electronics is that it dramatically and permanently changed the way we interface with computers.

If you are under 50 years old, you probably have never had to deal with a computer or any other digital device that doesn't have a graphical user interface. But computers prior to the Mac had a command line interface. You type in a command, the computer responds. Then type in another line and get another response. But you needed to know the commands, which can be like learning another language. The most efficient command line structures were also the most complex and hardest to master. In comparison, the point and click approach with a graphical user interface is intuitive and much easier to learn and master.

The history of the Mac started a few years before its introduction. For several years before the Mac, various research groups in Silicon Valley were trying to find an improved computer interface. In late 1970 for example, a guy named Douglas Engelbart at Stanford Research Institute invented and patented the mouse point and click device. It was first used on a Xerox Alto Computer system in 1973.

At Xerox's PARC (Palo Alto Research Center), Alan Kay and others were experimenting with graphical user interfaces. In 1981, Xerox introduced a software package called Star for its 8010 workstation which could possibly be the first graphical user interface in production.

In the 1981-1982 timeframe, Apple was already working on new computers to replace the Apple II and compete with the IBM PC. There was the Lisa (named after Steve Jobs' daughter), which was a very high-end computer based on the Motorola's 16 bit 68000 processor. There was also a lower cost version named Macintosh which originally was supposed to use Motorola's high-performance 8 bit 6809 processor.

Steve Jobs was very interested in the stuff going on at Xerox PARC. He was given tours and allowed to see all the cool stuff going on there like mouse controls and graphical user interfaces. He seemed to be the first person to fully grasp the potential of these new technologies for application in personal computers.

Back at Apple, Jobs quickly started to alter the Lisa and Macintosh products to incorporate these new technologies. He also hired several people from Xerox PARC to help implement these technologies at Apple.

The Lisa came out first in 1983. While it was a good trail run for the mouse and graphical user interface, it was vastly overpriced for the personal computer market at $10,000. It was a commercial flop. And even with the high performance of the 68000 processor, the graphical user interface put very high demands on the processor and made the Lisa seem slow and unresponsive.

So even before the Lisa was introduced, the decision was made to upgrade the Mac and use the 68000 for it as well. A mad scramble was on at Apple aimed at cost reducing the Lisa technology and then incorporating it into the lower cost Mac.

In trying to keep the cost down, a number of trade-offs needed to be considered. As usual, one of these was the amount of RAM. They settled on 128K, but it turned out to be a serious limitation and the subject of a lot of criticism once the Mac was introduced. Another compromise was the monitor screen size (9 inch), the number of pixels (512 x 342), and the color scheme (black and white).

The Macintosh 128k was announced to the press in October 1983 and was introduced in January 1984. It came bundled with two applications designed to really show off its new interface: MacWrite and MacPaint. It sold for $2500, which was still pricy, but not nearly as out-of-reach as the Lisa.

The reaction of computer enthusiasts to the Mac was huge. The Mac was a sensation, but it did not take long for people to start noticing that additional software for the Mac operating system was very scarce. It was not easy to program applications for the MAC. Redesigning existing applications that ran on command line systems was difficult and few people even knew how. But other software eventually came: Microsoft's Multiplan spreadsheet software came out for the Mac in April 1984. Microsoft Word came out for the Mac in January 1985. Lotus Software came out with Lotus Jazz, a spreadsheet and word processor suite, later in 1985.

So now there was at least some software for the Mac, but its limitations were becoming very obvious. The 128K bytes of RAM were very limiting, as was the lack of any easy way to expand its capabilities, i.e. no extra card slots like on the PC. It also did not have a hard drive available as an option.

Sales of the Mac slowed after some great initial numbers, and by the end of 1984, Apple was left with large inventories and declining profits. This brings us to another tale of intrigue from inside the personal computer industry.

In 1984, just before the introduction of the Macintosh, Steve Jobs had hired John Sculley away from Pepsi to be the CEO. Sculley was the president of Pepsi and was reluctant to leave. Jobs famously convinced Sculley to join him at Apple with the line: "You want to sell sugar water for the rest of your life, or do you want to come with me and change the world?" At the time, Jobs was 28 and Sculley was 44. It was common in

Silicon Valley for young company founders to bring in older, more experienced executives to help them run their businesses. Steve was content to focus his own efforts on running the Macintosh division of Apple. But by the beginning of 1985, Jobs and Sculley were disagreeing on many issues. Not only that, the disappointing sales numbers and lower profits were causing layoffs at Apple.

In April 1985, Sculley asked the board to remove Jobs from all operational control in the company. He wanted to limit Steve's role to chairman and public visionary for the company but nothing more. At the same time, Jobs was asking the board to remove Sculley. After two days of meetings, Sculley won. A few months later, Steve Jobs left Apple to found NeXT Computer.

Subsequent versions of the Mac quickly addressed the limitations of the original Mac128. In 1985, the Mac512 increased the RAM to 512 Kbytes. The Mac Plus was introduced at the beginning of 1986. It had 1 Mbyte of RAM, and a SCSI port through which hard drives, scanners, and other peripherals could be interfaced with the Mac. Then in 1987, Apple introduced the Mac II, which had expansion slots like the IBM PC for peripherals and other options. It also used the 68020, a 32 bit version of the Motorola 68000 processor. And with the addition of a video card, the Mac II supported 256 color graphics on a 640 x 480 color monitor.

During the 1980s, the Mac enjoyed a lot of success. The popularity of the graphical user interface and mouse made the Mac operating system seem vastly superior to the MS-DOS of the PC. Needless to say, the Mac left Microsoft scrambling to come up with its own graphical user interface. By 1985, they introduced Windows, followed by Windows 2.0 in 1987 and Windows 3.0 in 1990. All three were basically extensions of MS-DOS which added graphical interface functionality. But Windows initially failed to be as good as the Mac. It wasn't until Windows 95 in 1995 that Microsoft and the PC finally had an operating system that could compete with the Mac.

Some critics of Apple have pointed out that for an entire decade from 1984-1994, Mac had a much superior operating system, and yet, in spite of some nice success, could have done much better. Had Apple worked harder at reducing costs, and then priced Macs to be competitive with PCs, perhaps Apple could have emerged with a much bigger share of the personal computer market. Instead, they believed people were willing to pay more for a superior product and kept their prices well above those of PCs. Once Windows 95 came out, the big opportunity to gain market share against the PC was lost.

While Mac fans have always thought the Mac OS is superior to Windows, once Windows 95 was introduced, Windows could do almost everything that the Mac OS could.

My First Experience with a Mac

When the MAC first came out in 1984, I knew I had to have one. I read all about it a few months before the introduction. It was totally different than all my other experience with computers. I signed up on a waiting list at a local computer store. There were no Apple Stores then. I ended up getting my MAC 128 within a day or two of its introduction.

The MAC used Motorola's 68000 microprocessor, had 128K bytes of RAM, and a 400K byte 3 ½ inch floppy disc drive. Everything including the built in 9 inch monitor was in one integrated compact package. It cost $2500, which was a lot of money compared with that amount today. But I thought it was amazing compared to anything else available.

As mentioned previously, the Mac came with MacPaint and MacWrite applications built into the operating system and stored on the internal ROM. The Mac had a serial port with which you could interface it to a printer, so it was fairly easy to print out the results of your efforts with MacPaint or MacWrite.

I guess my initial experience with the Mac was like everyone else's. Although Mac Paint was just a plain paint app, and MacWrite was just a simple WYSIWYG (what you see is what you get) word processor, they were completely new in 1984. It was fun to show your friends, and everyone was amazed. But after the first few months, the lack of other applications became very obvious and very frustrating. Also, the 128K bytes of RAM was a frustrating limitation. You were always getting error messages due to lack of memory. And once you started generating lots of output from your MacPaint and MacWrite apps, that little 400K byte floppy disk was pretty annoying as well.

A year after the Mac128 was introduced, an upgrade package became available which increased the RAM memory from 128K to 512K, and added a 10MB internal hard drive. You had to give up your computer for a few days, because you couldn't install the upgrades yourself.

I had to get the upgrade too! But that was another $2500! So I then had a really cool computer, but for $5000 you could also buy a really nice car back then!

I remained a big Macintosh fan for many years. I was a big advocate for Macs, and even talked a lot of other people into buying later generation Macs.

I don't even remember how many other Mac models I've had over the years, but I know I had more than one generation of the Mac II.

I do have one regret about my original Mac 128. The daughter of a friend was starting college in roughly 1990. She was in bad need of some kind of computer. I gave her my Mac 128. It is now a real classic – kind of wish I'd hung onto it.

More on PCs and Microsoft

As we stated in a previous chapter, Apple's Mac left Microsoft and the PC industry scrambling to compete with the Mac's superior operating system. But business was mostly sticking with the PC, even during the decade from 1984 – 1994 while Microsoft Windows was struggling to compete with the Mac OS. There are several reasons for the success of Microsoft/Intel based systems. We will spend some time in this chapter talking about them.

Open Architecture – Hardware

We talked about IBM wanting to get the PC out quickly and using industry standard parts to produce it. Lack of any proprietary hardware made the PC easy to copy or clone. But the PC was also an open box – i.e., it had a removable cover with a bunch of empty expansion card slots inside. This meant that any company could build a card for the PC that gives it more memory, better video, better performance, or added functionality. The combination of multiple PC companies and a myriad of expansion cards gave the PC an incredible numbers of competitors with thousands of options as to features, capabilities, and options. The market loved this about the PC. The options were endless, and intense competition kept prices low.

But having an open hardware architecture is a double-edged sword! It creates problem for the user, for the maker of peripherals, and for the operating system (our next topic). With an infinite number of available options, every PC is a

---

little different. If you're making a new expansion card for the PC, you don't know what's already there and whether it's going to compete with your card for resources or perhaps prevent it from working at all. For users, anyone who has ever installed new hardware on a PC knows about the problems. Installations are sometimes plagued with problems. Sometimes new add-on hardware has to be returned, because it's simply incompatible with your particular system.

Open Architecture – Software

To accommodate all the possible hardware options, the PC operating system, whether it's the original PC-DOS or the newest Windows 10, needs to be capable of accommodating all the hardware variations. To do that it needs to be sufficiently open and flexible. It needs to provide access so that both external hardware designers and application program designers can get into the operating system and modify it enough to accommodate the requirements of the new hardware or new application. Again the market loved this open approach. It allowed the PC to adapt and be the solution to every problem that could benefit from automation and the application of computers.

But having an open software architecture, just like open hardware architecture, is a double-edged sword. It makes the operating system vulnerable. It means computer viruses can be introduced. Hackers can get into the system and monitor your activities. And with numerous different software companies adding stuff into and onto the operating system, unforeseen and unintended conflicts arise. Things get messy and unpredictable.

Price

Besides an open architecture, the other big factor in favor of PCs was their price. Both business and consumer customers found the Mac to be expensive even if you loved it. Everything for the PC was cheaper, both because of the much higher volume and because of the large number of competitors fighting over the PC market. In a very rough approximation, doing things the Mac way cost about three times what doing things the PC way cost. That was true in the 1980s and it's still roughly true today. So price was also a big reason for the dominance of the PC. You had to really love your Mac to justify the extra cost.

So price and open architecture, both in hardware and software, made PCs successful against the Mac. Apple kept their prices high, thinking people were willing to pay a premium to get the best. And Apple took great care to keep their hardware and software proprietary, and to limit how much either one could be modified. On the plus side, it meant that new hardware or software installations usually worked the first time. And it made the Mac much harder to hack or infect.

In today's world where we all have to worry about cybersecurity on a daily basis, you might think this would have been a great advantage for the Mac. But cybersecurity wasn't a huge issue back in the 1980s and 90s. Open architecture gave the PC a tremendous advantage in the marketplace back then! The PC held onto its dominant market share throughout this timeframe.

The other thing we need to discuss in this chapter is Microsoft Windows. Bill Gates was aware in the early 80s of what

Apple was doing with the Lisa and Mac. And he knew they had a big head start with graphical user interfaces, windows, mouse navigation, etc. So about the same time Apple introduced its Lisa, Gates announced Microsoft Windows in November 1983. It was just an announcement though. The first version of Windows wasn't actually available to purchase until two years later, in November of 1985.

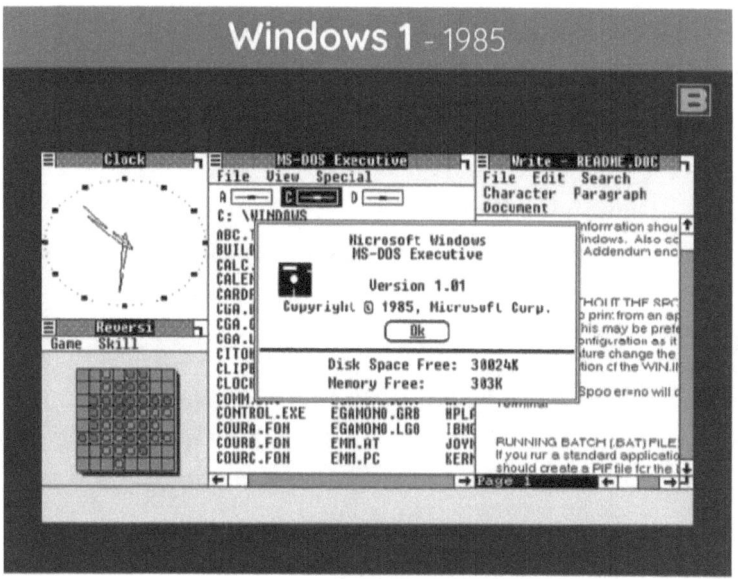

Windows 1.0 was a $100 add-on to PC-DOS. It provided tiled windows in which applications could run, providing a limited version of multi-tasking to MS-DOS. In addition to introducing PC users to the mouse, it came with a few sample programs: Paint, Write, Notepad, Calendar, Calculator, and the game Reversi.

Windows 1 was greeted by the market with mixed reviews. Users were delighted that the PC now had a graphical user interface like the Mac. They knew that Windows was the way to go for the future but were a little disappointed that this first version didn't do more.

Two years later, in December 1987, Microsoft introduced Windows 2.0. Like Windows 1, it was a $100 add-on to MS-DOS. Windows 2 allowed windows to overlap, as well as be minimized and maximized. It also introduced the Control Panel for controlling system settings. Microsoft Word and Excel for Windows were introduced for the first time. And Windows 2 required a hard drive for the first time.

Windows 2 did a better job than Windows 1 at getting Apple's attention. They sued Microsoft for trying to steal the "look and feel" of the Mac. Almost every aspect of the suit failed, so Microsoft was now free to make Windows as similar to the Mac as they wanted.

Windows 3.0 was introduced in May 1990. It was a full operating system in its own right, not something augmenting MS-DOS. It allowed MS-DOS programs to run in windows, giving MS-DOS its first real multitasking capability. It had a Program Manager that is very similar to what we find still today on PCs. It also introduced PC users to Solitaire and Minecraft. Windows 3 generally received better reviews than Windows 1 and 2. Windows 3 was finally showing that the Microsoft operating system was starting to compete effectively with the Mac.

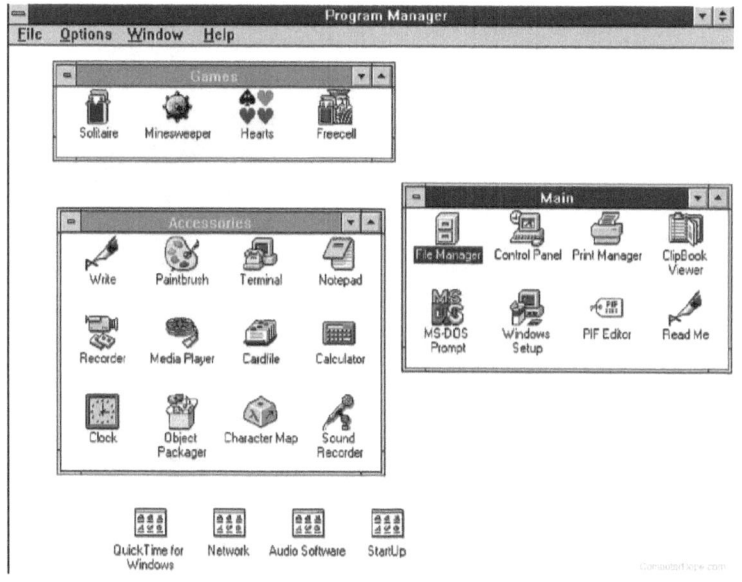

Windows 3 Program Manager

And finally, in August 1995, Microsoft introduced Windows
95. Windows 95 introduced a whole host of things we still use
today and think of as basic to Windows, such as the task bar
and the Start button. It also introduced true preemptive multi-
tasking. Windows 95 also introduced "plug and play", where
Microsoft made a good effort to streamline the installation of
peripheral devices by automating the installation process,
finding the needed drivers, etc. While it didn't always work, it
was pretty impressive when it did.

With Windows 95, the desktop became the place it is today for
Windows, where we place shortcuts to applications, photos,
etc. In previous versions, the desktop served only as the
location of windows or icons for running applications.

By the time Windows 95 was introduced, Internet support was starting to be of interest.   We are going to discuss the Internet shortly in an upcoming chapter. The Internet would soon change everything. But for now, we'll just mention that Internet Explorer appeared as an application for the first time with Windows 95.

One last thing worth discussing in this chapter is what happened to IBM in the personal computer industry. As we said previously, the IBM PC made Intel and Microsoft the two big powerhouses in the personal computer business. IBM was relegated to being just one of many companies making PCs. By 2004, PCs had become truly a consumer electronics business. Consumer products wasn't really a business IBM wanted to be in, and it wasn't particularly profitable for them either. In December 2004, they announced they were selling the PC business to the Chinese company Lenovo for $1.75 billion!

Motorola Part II

After several years in Thyristors, I wanted to do something else. My next job was something very different. I was hired as Manager of Strategic Planning for the Semiconductor Division of Motorola. It was very different because I had been in operations, but this was a staff position, and it applied to the entire semiconductor business of Motorola.

I had a number of different responsibilities in this job, but the biggest one was getting the roughly 25 operations that made up the Semiconductor Division to each put together a 5 year strategic plan, in a consistently organized manner, and then to consolidate those plans into a 5 year strategic plan for the entire Semiconductor Division.

The problem was each operation had its own ideas about how fast it could grow, how many people it needed to achieve that growth, how much capital investment it needed etc. But then our division manager and his staff had their own ideas about what those numbers should look like for the entire division, and the corporation had its own ideas about what the semiconductor plan should look like. The various ideas were always wildly divergent. If you left it up to the operations, they all planned to double in size every two years, but needed 50 times more capital funding than Motorola corporate was willing to invest.

So my job was to manage the up/down gyrations of these plans, until we had a consistent set of plans that added to something that the division and corporate management were both happy with. It really did involve a lot of numbers, because not only were there 25 different operations that made up the semiconductor division, but the plan went out for 5 years with the first two broken out by quarter.

That's where my Xerox Sigma 9 computer programming came into play. I used it to manage a running total of the sums of all the individual plans as we tried to get all the pieces to fit together into a sensible plan for the total semiconductor division.

I learned a lot of new stuff in a very short time in this job. One reason was that I was surrounded by a lot of very, very talented people. I felt a little out of my league for the first few months. My boss was a brilliant expat from France named Pierre Lesieur. After I had worked for him for about a year and a half, he decided to return to France. To my surprise, I was the one chosen to replace him, so now I was the Director of Planning for the semiconductor division.

This job involved some other functions besides strategic planning. It included some economic modelling, some financial planning, and managing the capital investment planning and allocation for the division.

The capital investment planning was particularly interesting. This was a strange time. I already mentioned in a previous chapter that semiconductor manufacturing was becoming super high-tech during this period. But super high-tech also means super expensive. A wafer fab where the silicon wafers are processed into chips cost maybe $2 million in the 1960s. A fab like that could make discrete devices or simple integrated circuits. But by the mid-70s, the investment cost had increased

1000 fold. Individual pieces of equipment were now $2 million, and a state-of-the-art MOS integrated circuit fab now cost $2 billion.

The technology was changing rapidly, as integrated circuit geometries got smaller and crammed more and more transistor elements onto a single chip. It was normal for a major player in semiconductors to have multiple IC fabs under construction with different capabilities and perhaps on different continents, so Motorola Semiconductor was suddenly asking the corporation to invest $10s of billions every year. It created an uncomfortable strain in relations between semiconductor and corporate. Motorola couldn't come up with that kind of investment without borrowing money and starving its investment in its other businesses.

About this time, I got to know a man at Motorola who made a big difference in the company. His name was Al Stein. He came from Texas Instruments and was hired by Motorola to run the Integrated Circuits Division. (At some point about now, Motorola had promoted its semiconductor business to a Group, with separate divisions under it.)

At the time Al arrived, we were really struggling in integrated circuits. We weren't cost competitive; we were behind others in many of the key technologies; we weren't investing fast enough. And as a result of all of this, we were losing market share.

Al arrived and wouldn't have any of this. He found what he considered a complacent attitude across the whole organization. He demanded change and made everyone's life miserable. He had his staff and staff's staff all working 16 hours a day. He was totally unhappy with the support he was getting from the management above as well.

This was a time before cell phones and before everyone carried their laptop with them everywhere, so Al was quite unusual by demanding everyone's home phone number. Al only had about 8 or 10 direct reports, but his management style was to attempt to directly manage at least 100 people. It didn't matter if you worked for him directly, indirectly, or didn't work for him at all. You would get calls from him at 9:00 at night or 6:00 in the morning, asking you to fix something or get him some information.

There was a general knowledge that everyone disliked Al and considered him totally unreasonable. If you came to a meeting without something he was expecting, or came in any other way unprepared, he made a great point of throwing you out of the meeting and telling not to return until you were fully prepared. He seemed to particularly enjoy doing this to his own staff members.

This went on for a long time – perhaps a year and half. We all wondered if there might be a full scale revolt as a result of Al being so demanding. But then something strange starting happening – the Integrated Circuit Division started making money. It started gaining market share. Al Stein had pretty much turned it around single-handedly! Everyone in the IC division suddenly felt like a winner, and Al's demanding management style was all of a sudden a lot easier to tolerate.

My own relationship with Al Stein was interesting.   On a couple of occasions I was thrown out of a meeting for not being as prepared as Al thought I should be. So I certainly shared everyone else's frustrations with him, but I was also very fortunate. I became one of the few people he seemed to like and trust. He took me with him to a meeting in New Orleans. I don't think I was much help to him, but he wanted me there as an advisor. Al gradually started to respect me. He started asking for advice on things I felt completely unqualified to advise him on. So in the end, I had a good relationship with Al.

Al Stein, after he left Motorola and became CEO at VLSI

Al Stein was promoted to assistant general manager of the Semiconductor Group and I continued to work with him in that capacity, but he remained a polarizing figure, antagonizing a lot of people above and below him. He finally left Motorola to become the CEO of VLSI Semiconductor, and then later CEO of distributor Arrow Electronics.

Another interesting thing was going on about this time, both at Motorola and its competitors. Product quality had always been an issue with semiconductors. One silicon wafer contained anywhere from a few hundred to a few thousand devices. Some were good while others were defective. The more complex the device was (or the more demanding its specifications were), the lower the yield of good devices per wafer.

Devices were tested while still in wafer form and bad devices were marked with an ink dot. Inked devices did not get packaged. So there was a "yield loss" at wafer probe. After the good devices were packaged, they were tested again. Some failed final test, so there was an additional "yield loss" at final test. It was not uncommon for total yield losses to be significant; perhaps only 30% of original devices on a wafer turned into good finished devices. Then at the customer's assembly line, additional defective devices were commonly found. Some were damaged by the customer. Others were defective to start with but somehow escaped all the testing and ended up in a box of supposedly good parts. It was just the way things were; the whole industry put up with poor yields and a percentage of defective parts reaching the customer.

The Japanese had for many years had a reputation for poor quality products, not just semiconductors but everything manufactured. But suddenly things started to change. Semiconductor customers started buying products from Japan because their quality was better than that of U.S. suppliers.

Japanese semiconductor suppliers started taking market share away from American suppliers. At the same time Japanese car manufacturers started successfully selling large numbers of cars in the U.S. for the first time. The reason was the same – significantly better quality control.

Our Motorola Semiconductor Assistant GM, Bob Heikes at that time, posted huge banners at the entrance to each of our plants: "The Japanese Are Coming!" it said. We had to start fixing our quality problems.

Ironically, we soon found out that American quality experts like Phil Crosby had been spending time in Japan teaching Japanese companies about quality control, in part because they hadn't had much luck getting American companies to listen.

Crosby's big tagline was "Quality Is Free!" For the semiconductor industry, that meant that if you could get those yield losses drastically improved, you would not only get a lot more saleable devices out of one silicon wafer, but you would also would have far fewer defective ones with a chance of escaping all your testing and ending up in a customer's circuit board. Crosby maintained that even significant cost expenditures to fix your quality problems would be offset by even more significant cost improvements.

This quality challenge very quickly became a big initiative at Motorola Semiconductor. It quickly spread to the rest of the U.S. semiconductor industry. It also spread throughout Motorola. It also spread across the U.S. automobile industry.

One particularly significant result of all this new emphasis on quality was the Six Sigma Program. Six Sigma is a statistical term that suggests less than 3.4 defects per million or a 99.99966% yield. It's the kind of stuff Crosby meant when he said you should strive for "zero defects". The Six Sigma Quality Program was invented by Motorola, and quickly spread across American industry. It earned Motorola the first coveted Malcolm Baldrige National Quality Award in 1988.

In general we have seen huge improvements in the reliability and quality of American manufactured goods since the 1980s. Electronic consumer products and automobiles are perhaps the two most obvious things, but the quality of all manufactured goods has improved significantly over the past 40 years as a result of the quality initiatives started at Motorola in the 1980s.

I will finish this chapter on working at Motorola with one more story. I worked as Director of Planning for about 5 years, but it was a staff position and I really wanted to get back into operations management. I moved into a position called Director of Materiel in Motorola's Power Products Division, and then later, into the same position in the Linear Integrated Circuits Division.

These positions included managing a number of business functions: manufacturing planning, materials procurement planning, and customer service - basically all the logistical support for a large manufacturing operation. It was also a catch-all for other support operations when they didn't fit anywhere else. So part of the time I had a final test operation, a product warehouse, plant facilities, industrial engineering, information systems, and other miscellaneous functions reporting to me. This was the kind of job I found myself in as personal computers came into widespread use and changed the way everything was done in our business.

Each division at Motorola Semiconductor produced anywhere from 1,000 to 10,000 different products, each requiring a specific silicon chip, a package frame, various assembly components, and a test program that ran on a specific test system. The silicon might be fabricated in any one of several wafer processing fabs in the U.S, Europe, or Asia. It might be packaged in any one of several assembly/test sites in Asia. And that specific product might have customers ordering it all over the world. Managing all this before computers was an almost mind-boggling system of manual paperwork. We had to decide how much of each thing to produce each week, where it was to be done, where it was going to go next, etc.

For years, the only computer support for all this was a report generated every morning called the Daily Status. It showed for every device, where the inventory was by stage of completion (wafers in process, die inventory, parts in assembly, finished devices in inventory, etc.), and all the time-framed open orders (delinquent, due next week, the week after …etc.). Then based on that daily status, all kinds of manual reports and schedules were devised, and based on those came factory run rates, test schedules, etc. Once personal computers started to appear in the workplace, these scheduling and manufacturing planning functions were one of the first to be revolutionized by personal computers.

At first it was spreadsheets that completely replaced all the manual worksheets. Then various planning systems were devised which collected inputs from multiple organizations and produced consolidated factory schedules. With all this came enormous improvements in productivity!

But the arrival of personal computers changed a lot more than just the stuff that desperately needed to be automated. Email changed everything about communication. Before email, written communication was expensive and used only when necessary. Large distributions were also difficult, so everything was on a sort of "need to know" basis just because of the cost.

In the 70s and early 80s, Motorola Semiconductor was already an international business with factories all over the world. But only the highest level managers talked on the phone or actually visited our international factories and offices. Email totally changed that. Suddenly we had people at all levels of the organization working directly with their international counterparts.

The world shrunk significantly! Learning, shared information, international cooperation, and global productivity soared. Within 10 years after the arrival of personal computers and email, it was already hard to remember how we got anything done without them.

Networking and the Internet

So far we have only talked very briefly about the Internet. We mentioned the impact that email has had in the last chapter. Email is a product or service of the Internet. We also mentioned in a previous chapter that, with the arrival of Windows 95, the web browser Internet Explorer first became available. But from here on out, it is increasingly difficult to talk about digital electronics without talking about networking and the Internet.

In the early days of computers, they were seen as super-fast computational machines that could automate bookkeeping and accounting. But when we step back and ask how computers have changed the world, the answer has more to do with communication and an enormous increase in our access to information. And that has more to do with the interconnectivity of computers than to the nature of an individual computer.

In this chapter we will explore the basics of computer networking and the history of the Internet. By the way, just as a point of information, when Internet is capitalized, it refers to the one global Internet. Local internets, usually referred to as intranets, are smaller networks, perhaps within a company, university or government, which may or may not be connected to the global Internet.

So where did this all start? Networks at a local level, known as LANs (local area networks), started to appear almost as soon as computers themselves. But they did little more than

---

allow multiple users to connect to one computer, or a small number of computers to exchange files. There were no standards; each LAN might have its own set of protocols. That didn't really solve the issues of connecting multiple computers over large distances.

Well before the emergence of personal computers, computer engineers recognized the value of and need for interconnecting large numbers of computers over long distances. They saw the need for wide-area networks or WANs. In the early 1960s, people started proposing methods for routing packets of data between computers. The first actual implementation of this was called ARPANET, or the Advanced Research Projects Agency Network. It was funded by the U.S. Department of Defense and connected computers in several universities during the late 1960s. UCLA and SRI (Stanford Research Institute) were the first two, followed by UC Santa Barbara and University of Utah. Many additional universities were added over the next couple of years.

During the 1970s, work continued to make ARPANET function effectively. It was viewed as a network of networks. The need for standards that could be used to connect multiple local networks together became obvious. Two standards were adopted for this purpose: Transmission Control Protocol and Internet Protocol, or TCP/IP. Together, these two protocols formed a communications model for transmitting data between multiple networks. They provide the basic functionality we see in the Internet today: the ability to identify a specific computer by its IP address, and the ability to route data packets between specific computers. ARPANET adopted TCP/IP in

1983, and from there it began to assemble into what has become the modern Internet.

Robert Kahn

A large number of people contributed to ARPANET and later the Internet, so it's difficult to point to specific people as inventors of the Internet. But Robert Kahn, who was a co-inventor of the TCP/IP protocols, deserves some special mention. He proposed the ground rules that shaped the nature of today's Internet:

1. Each distinct network would have to stand on its own and no internal changes could be required to any such network to connect it to the Internet.

2. Communications would be on a best effort basis. If a packet didn't make it to the final destination, it would shortly be retransmitted from the source.

3. Black boxes would be used to connect the networks; these would later be called gateways and routers. There would be no information retained by the gateways about the individual flow of packets passing through

them, keeping them simple and avoiding complicated adaptation and recovery from various failure modes.

4. There would be no global control at the operations level.

The last item, "no global control", is worth talking more about. The Internet is not run by some big company. It is just a bunch of rules by which computer networks can communicate with one another. There does have to be companies providing the physical infrastructure to make that happen. They're called Internet Service Providers or ISPs. But there are thousands of them, and no one company controls the Internet. I know that some people will contend that Google and Facebook control the Internet. We will talk more about this a little later in our chapter about cybersecurity, but conceptually, the Internet is an open architecture with no central control structure.

In the days of ARPANET and the early Internet, use of it was limited. The general public barely knew of its existence. Various uses included file transfers between computers using the FTP protocol, chat room applications using the IRC protocol, and of course email. There were many others, but we won't try to discuss them here. The use of the Internet quickly grew and its architects at ARPANET had to make rapid modifications as they started to think about hundreds of millions of users. An example of these modifications is domain names. Originally, computers were simply identified by their IP address, but with millions of computers on the network, people needed an easier way to identify a specific computer. So domain names were added to the Internet. A new piece of hardware infrastructure – the domain name server, was also added, which translated domain names into their respective IP addresses.

Tim Berners-Lee

But in the 1990s, everything changed for the Internet. A new protocol came onto the scene. It was called Hypertext Transfer Protocol or HTTP. HTTP and the World Wide Web were proposed and first implemented by an English computer scientist named Tim Berners-Lee. Tim created the first website server on a NeXT computer at CERN in Europe. He also created the very first web browser, which he simply called WorldWideWeb. Although websites started out as simply a collection of text pages interconnected by clickable hyper-links, they changed rapidly with the addition of images, video, audio, and a host of other features.

Another major milestone for the Internet and the World Wide Web occurred in 1993, when Marc Andreessen, who later founded Netscape, launched the Mosaic web browser. Mosaic and then the Netscape browser became the models for all modern web browsers. These browsers were what made the Internet and websites popular with the general public. They made the Internet easy to use for everyone. Then in 1995, Microsoft introduced Internet Explorer. From then on, everybody was using the Internet for everything.

Today we tend to use the terms Internet and World Wide Web pretty much interchangeably. But just to reiterate, the World Wide Web is just a particular protocol running on the Internet. It happens to be the one we use every day, but it is not the same as the Internet. The Internet is the global wide area network that has resulted from the standards and protocols originally set by ARPANET.

As we continue to talk about the Internet, we quickly get into all kinds of grey areas – where we worry about the security of Internet commerce, how the Internet compromises our privacy, whether our identity can be stolen, whether hackers can steal our money or pictures, or unlock our front door. We will talk some about all of this in a later chapter on cybersecurity.

If You're Interested:
How Websites Work

As a preface to this chapter, I should probably mention that I started a small website design company after retiring from Motorola. It was called D2 Web Design. Some of the materials in this chapter were designed to show my customers how websites work.

The Internet and websites are amazing technologies that we all take for granted today. But how does the Internet actually work? In this chapter, we'll first look at the big picture – what is the Internet, how do the major pieces work, and then, specifically, how do websites work. Then we'll look at how websites are created.

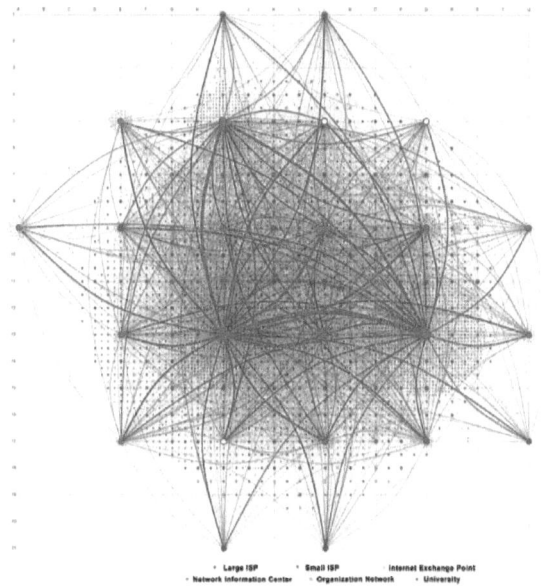

Let's start with the really big picture first. The Internet is an incredibly large network of interconnected computers. Hundreds of millions of these computers have users viewing web pages or reading emails, while millions of other computers are servers supplying the webpages or distributing emails.

The amount of information being exchanged on the Internet is staggering. By the way, notice that Internet is capitalized. There are local intranets, company's internal internets, etc., but there is only one global Internet. The capital I signifies "The Global Internet"!

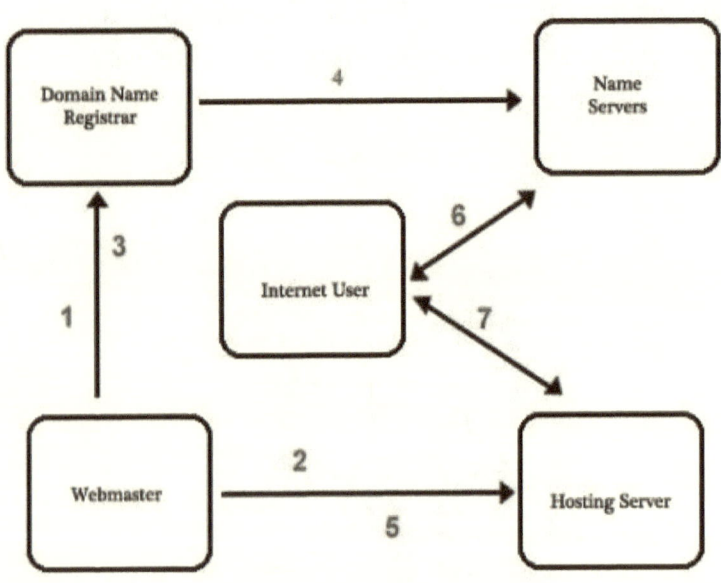

But how does this all work? What are the major processes at work that allow us to see websites? Let's take a look at how a website is created and put on the Internet. First, we will *examine how domain name registration, website hosting, and website design all fit together to create websites. After that, we will show you how the various files that make up a website come together to create website pages.*

The diagram above shows the various systems that work together to create a website and make it accessible to the public. Each box represents a different computer or server, all interconnected across the Internet. Let's first define each of these systems and then discuss each interaction between them.

The Domain Name Registrar is a company capable of registering a domain name, such as abc.com or xyz.org. They assign ownership to that name. The best known registrar is GoDaddy.com.

Name Servers are servers distributed throughout the Internet that translate a domain name into the physical address (called an IP address) of the server on which that website resides. Your own Internet provider will have a name server, so when you go looking for a website, this server can tell your web browser where to find it.

Internet User is just as it sounds - the end user who is viewing a website.

Webmaster is the person who creates and maintains the website.

Hosting Server is the publicly accessible server on which the website is stored and made available to the public. Hosting servers are not much different than a regular PC; the main difference is that they are connected to the Internet with a very high speed connection, called a T1 line, capable of uploading large amounts of data at very high speed.

Now let's examine how these various pieces interact with one another.

Step 1 - the webmaster logs into a domain name registrar, selects an available domain name, and registers it. The webmaster needs to retain access to the registration account, as he will later come back to the domain registration, and add the name server codes for the hosting server.

Step 2 - the webmaster creates a hosting account for the new website by logging into a hosting service. He obtains login credentials with which he can transfer the website files to the host server using FTP (File Transfer Protocol).

Step 3 - the webmaster now adds hosting server codes to the domain registration. This provides a link associating the domain name with the host server.

Step 4 - the domain registrar distributes the domain name and associated hosting account to all name servers. Updating this information to all the name servers in the world takes several hours, so linking the domain name and the host account is not instantaneous. However, within a few hours, any user computer worldwide will be able to type in the domain name and be directed to the correct hosting server.

Here is our diagram again, just to keep it in view:

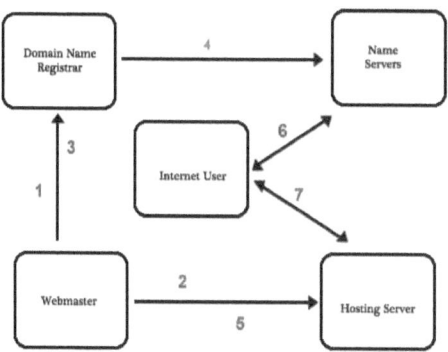

Step 5 - the webmaster can now build the website. The webmaster normally creates the website and all its component files on his own computer, and then uploads it to the hosting server when ready to be viewed by the public.

All of the above only happens once. We are now ready for any user to view the website. The next steps happen every time someone views the website.

Step 6 - the user types the domain name into a web browser. The web browser goes to the nearest name server to translate the domain name into the IP address of its host server.

Step 7 - the user's web browser calls the IP address of the host server and requests the main file (known as the index file) for the domain name.   The index file will in turn tell the browser what other files are needed to display the home web page.

Now let's look into what actually makes up a website. Some people assume that a web-page must be like a Word document, that is, a single file which is downloaded from a server and displayed on the user's monitor. Actually, it is quite different than that. A single web-page is typically made up of dozens of files, downloaded to the user's computer, and then built up into a web-page "on the fly" by the user's web browser. We'll now look at what those various files are and how they fit together to make a webpage.

The diagram below shows you some of the files that might make up a typical web-page. An HTML file is downloaded first to initiate the process. (HTML stands for HyperText Markup Language. It's the most common programming language used to create web pages.) In the diagram, labelled as item 1, is an HTML file called index.html. The index file is the default first page to be downloaded, so it is typically the file that starts the process of building the Home page.

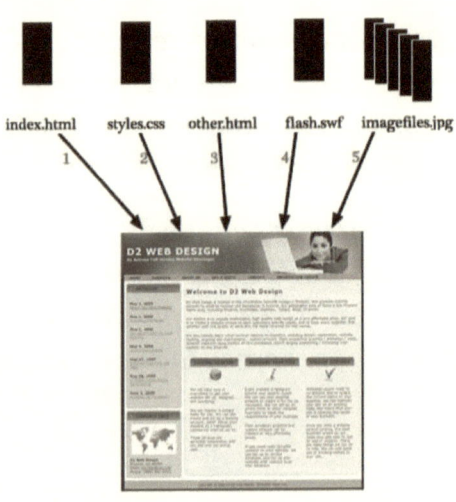

The index file provides all kinds of information about the page that will be assembled. It contains most of the text that will appear on the page. It can contain the header, the menu, the footer information, etc. But it makes numerous references to other resources (files) that will be used to assemble and display the page.

One of the other files referenced by the index file is the style sheet, labeled as item 2 in our diagram. The style sheet specifies the size, location, color, border, font, layout, etc. for each region of the page. By region of the page, we mean things like header, footer, left-hand column, right-hand column, menu area, etc. When a font is specified in a style sheet, it is specified as to size, preferred font, several substitute fonts to be used if the preferred one is not available on the user's computer, and font style (such as bold, underline, italic). The style sheet will also specify how a font will behave if it is part of a link - does it change color when the mouse is over it? Does an underline appear or disappear? Is the link a different color if it has already been visited?

Typically a single style sheet controls every page on the website. This has some great benefits. It guarantees a consistent look across the entire site. It allows the designer to change that look by altering a single file. And it separates the look of the website from the content of the website.

The index file may also call other HTML files as shown in item 3 in our diagram. For example, the webmaster may want to define a complex menu structure in a stand-alone HTML file, and then call it into the index file or other page files. That way, the menu structure is all contained in one place, even though it is repeated and appears on every page.

The index file may also reference a Flash animation, as shown in item 4 in our diagram. Again, each Flash element on a page is a separate file which needs to be downloaded.

Finally, all the images on a web-page are separate files, as shown in item 5 in our diagram. And typically, there are a lot more images on a page than you might expect. While a picture is an obvious image, many other elements of a web-page are also images. For example, anytime you want to completely control the way text looks, perhaps a specific font with no substitutions, or text with a shadow around it, the text has to be placed in an image file. Otherwise, the user's web browser will have the ultimate control of what that text looks like. Similarly, numerous small features on a web-page, such as separators, bullets, rounded corners, etc. are created with images. It is not uncommon for a single web-page to use 40-50 images.

So how does all this come together to create a webpage. Web browsers are so common today that we take them for granted, but they are really very complex pieces of software, capable of reading the instructions from that primary index file, downloading all the auxiliary files necessary to display the page, interpreting all the format instruction in the style sheet file, then assembling the whole thing and displaying it on the user's monitor.

With a slow dial up connection, you can sometimes see this process unfold the first time a web page is viewed, just because of the time involved to get all the files downloaded. With a high-speed connection, the page seems to appear all at once, but a whole lot is going on, right on the user's computer.

Another interesting thing about web browsers is that they anticipate having to use the same files again, either to display another page on the same site or perhaps the same page a few minutes later. All the files are saved for potential reuse - a process referred to as caching. Caching greatly reduces page loading times, as well as total Internet traffic. However, it also means you may not be seeing the most recent changes to a web page.

So now you know at least a little bit about how websites work.

More about Apple

In our chapter about the Macintosh computer, we talked about Steve Jobs departure from Apple in 1985, as he left to start a new company, NeXT Computer. We also discussed how Apple over the next decade had a clearly superior operating system compared to the PC, yet failed to gain market share against the PC. Then in 1995, that potential competitive advantage disappeared. Windows 95 came out and the PC had an operating system roughly as good as the Macs.

During Jobs absence from Apple, the company went through a lot of changes. At first, it had moderate successes with the introduction of full color Macs and the introduction of the PowerBook laptops. But then the Newton PDA was introduced and was poorly received. (We'll talk more about the Newton in a future chapter, when we talk about tablets.)

Next Apple entered an agreement with IBM and Motorola to switch processor platforms to the PowerPC. It was a significant change and required an expensive retooling of both the operating system and much of the Mac's available software. This change turned out to be a mistake, because the PowerPC, though it briefly looked to be superior to Intel processors, was much more expensive than Intel's x86 family of processors. Intel's processors had much higher sales volume, and they kept getting faster, better, and cheaper until the PowerPC was obviously unable to keep up.

Apple's profits were suffering badly from all this, and in June of 1993, John Sculley was forced out as CEO.

Sculley was replaced by Michael Spindler, a long-time Apple employee who held the CEO's job for three years. During this time, Apple engaged in acquisition discussions with IBM, Philips and Sun Microsystems, but all these deals fell through and Apple remained its own company.

In 1996, Spindler was replaced by Gil Amelio, who was a member of Apple's Board of Directors and had been president and CEO of National Semiconductor. Amelio got the idea of acquiring NeXT computer and bringing back Steve Jobs as a consultant. He was able to actually make it happen for $429 million in early 1997, though it quickly led to his own undoing.

Only a few months later in July 1997, Jobs convinced the board to fire Amelio and name him (Jobs) as interim CEO. He was able to convince the board to remove Amelio in large part because Apple stock had recently hit a 12 year low. A major cause of the stock price's new low was a recent anonymous sale of 1.5 million shares of Apple stock. After Amelio was safely out of the picture, Jobs admitted that he was the one who sold 1.5 million shares of Apple!

As interim CEO, Steve Jobs got right to work. He soon patched things up with his long-time rival Bill Gates, and convinced Bill to make a $150 million investment in Apple. He also correctly realized that Apple had too many different and confusing versions of the Macintosh. He significantly reduced the number of models.

Then in 1998, he and Apple introduced the iMac. The iMac was a success and revitalized the Apple brand. The combination of the reduced product line and the success of the iMac brought Apple back the brink of financial disaster.

The iMac in 1998

Steve Jobs was now viewed as the returning hero who had saved Apple. In January 2000, the interim title was removed and Jobs became the permanent CEO of Apple.

One benefit that Apple derived from its acquisition of NeXT Computer was NeXT's operating system called NeXTSTEP. Apple adapted this Unix based operating system to replace its Macintosh operating system and introduced it as Mac OS X in 2001.

The year 2001 was a big year for Apple for other reasons too! Apple started opening its retail stores in 2001. Apple also introduced the incredibly popular iPod that year as well. The iTunes app soon became the way to manage your music, whether you used a Mac or not. And then in 2003, Apple opened the iTunes online store. Five years later in 2008, iTunes had become the largest music vendor in the U.S. And by 2012, it sold 30% of all music sold worldwide!

iPod

In 2006, Apple finally gave up on the PowerPC processor and switched to Intel processors. It was a major change for the company. Intel processors were cheaper than the PowerPC processors, but in addition, Intel processors consumed less power and generated less waste heat.

We haven't talked yet about the iPhone or iPad, but we will in a later chapter. For now, let's just say that Jobs and Apple had a series of big successes, eventually leading to Apple passing ExxonMobil in 2011 to become the world's most valuable company.

History of the Cell Phone

Cell phones have changed everything. The phone used to be something that a family shared and took turns using. The idea of having your own personal phone number was strange. But today, we take for granted having your own personal phone. And the ideas that it's always with you and that you are always connected and reachable are things we also take for granted today. Being able to text message, having your contacts' phone numbers stored in your phone and many other features that we use every day were not available until the cell phone. So in this chapter, let's look at when and how cell phones came to be.

In 1947, William Rae Young, working at Bell Labs, proposed that radio towers arranged in a hexagonal pattern could support a telephone network. Young's design called for low-power transmitters carrying calls across the network. It also included handoffs from one transmitter to another as a caller moved from one transmitter's broadcast radius to another. All of this is how cell phone technology works today, but the technology to actually implement cell phones didn't exist back in 1947.

As we mentioned in an earlier chapter, mobile phones were around before cell phones. There were a few frequencies reserved for mobile phones, and operators manually connected these mobile phones to land-line phones. They were not particularly reliable. Service was pretty much limited to big cities, and disconnects were common. They were also very expensive. So phone engineers were looking for a better way - something that would make mobile phones less expensive and much more reliable.

By the early 1970's, electronic and computer technologies had progressed to the point where people at both Bell Labs and Motorola were trying to turn William Young's cell phone idea into a reality. As it turned out, an engineering team at Motorola led by Marty Cooper was the first to develop a working cell phone and demonstrate a working cellular network.

Marty Cooper with a DynaTAC phone

The phone itself was the Motorola DynaTAC. It was 9 inches tall and weighed 2.5 pounds – not exactly small, but it was the first hand held portable phone. It was first used by Marty Cooper on April 3, 1973 to call Joel Engel, his professional rival at Bell Labs. It would be another 10 years before the DynaTAC phone would go on sale to the public, mainly because it took that long to create and install enough cell towers and infrastructure to make commercial cell phone service practical.

Early cell phone technology, called 1G (first generation), was analog. That means that the voice transmission itself was not digital. The connection to the tower, the identity of the phone, and the switching of the call between towers were all digital and managed by computers, but the voice itself was carried on an analog transmission. That meant that just like early analog television, it was subject to interference and occasional cross-talk where another call might drift into your conversation.

Commercial implementation of cell phone systems actually happened in Japan before it did in North America. NTT got its cellphone system up and running for the city of Tokyo in 1979, four years before Ameritech would launch North American's first cellphone network using the Motorola DynaTAC phone on March 6, 1983.

The DynaTAC phone wasn't for everybody. In 1983 it cost $3995. It had a call-time of just 30 minutes and took 10 hours to recharge. But cheaper and smaller phones were soon to come. During the 1980s and early 1990s, Motorola was the dominant player in cell phones. They were recognized as the inventor of the cell phone and had most of the top selling phones. In 1994, Motorola had 60 % of the cell phone market in the U.S. They were also the top supplier of cellular infrastructure, i.e. the transmitters and computer systems that power the cell phone system.

MicroTac   StarTac

Motorola's MicroTac phone was introduced in 1989, and was the smallest and lightest at that time. Then in 1996, Motorola introduced the StarTac phone. The StarTac was one of the first phones to really catch on with consumers.

But in spite of all of Motorola's success, it had missed a critical market change that occurred in the early 1990's. In Europe in particular, cell phones were going digital. Digital cell phone technology, known as 2G (second generation) had all kinds of advantages over 1G analog. It allowed a slice of radio frequency bandwidth to handle up to 10 times more calls. It could transmit a call using much less power, which significantly increased cellphone talk-time and reduced battery consumption. It permitted encryption of calls for improved security. And it improved reliability it general.

Interestingly, Motorola's cellular infrastructure division knew the market was going digital and was building much of the equipment driving the transition. But at the same time, the handset division that actually made phones thought it would be many years before the bulk of cell phone business would go digital. Consequently, Motorola was not prepared for the rapid transition to digital.

The company that was prepared was Finland's Nokia. By 1997, Nokia had become the world's largest supplier of cell phones. It continued to hold the number 1 spot for the next 15 years. Motorola never recovered from its major strategic blunder; we will talk more about that later.

One other topic is worth mentioning at this point. There were (and still are) two different versions of digital cellphone technology. GSM, which stands for Global System for Mobile Communication, is used in Europe and throughout most of the rest of the world. CDMA, which stands for Code Division Multiple Access is the system used in the U.S. and Russia.

These are two competing digital technologies. The fact that there never was a clear winner between them may have contributed to Motorola's delay in going digital. Today we don't worry much about it. Many phones today are both GSM and CDMA compatible, meaning you can travel back and forth between the U.S and Europe and your phone works fine in both places.

So this gives you some background on the early history of cell phones. We haven't talked yet about how cells phone turned into smart phones, giving us Internet access, cameras, GPS navigation, texting, and all the other stuff we enjoy today. We'll talk about that in an upcoming chapter.

The Whole World's Gone Digital

In this chapter, we'll look at some of the products that have gone digital over the years. It's an amazing array of products. Many digital technologies can trace their origins to the early 1970s. That's when integrated circuits were first capable of containing a few thousand transistors. Suddenly processes and products that had always been analog could be redesigned to use digital technology at a reasonable cost. So at that time lots of products started to undergo redesign. Some like television took many years to go completely digital. But others happened very quickly. One we've already talked about is electronic calculators. So we'll start here with another one that happened quickly - digital clocks and watches.

Hamilton Pulsar

A clock or watch with an LED digital display and all digital timing mechanism could be built from a few thousand discrete

---

components in the 1960's. But in the early 70s, single IC chips containing all the necessary components appeared on the scene. National Semiconductors MM5316 was one such chip. Almost immediately, digital watches and digital clocks began to appear.

When first introduced in 1972, a digital watch like the Hamilton Pulsar was a luxury product for the rich at $2100 (which would be more like $12,000 today). But just like the hand held calculator, 10 years later, you could buy one for $5!

I remember reading an article, probably around 1980 when digital watches were becoming very inexpensive. It said that the goal of every human on the planet was to own a watch and a camera. And it implied that, with the way digital watch prices were coming down, it was becoming a realistic goal.

Well things changed. Digital watches went out of fashion, and more recently, as a new generation grew up with smartphones, watches of all kinds have gone out of fashion. (At least smart watches like the Apple Watch are still in fashion.) But the prediction that every human wants a watch and a camera was correct. Now every human wants a smartphone that includes a watch and a camera!

Another place where technology has turned to digital is audio recordings. Thomas Edison invented the phonograph in 1878. It used a cylinder to record analog music and sound. By 1940, vinyl records came out, and analog recordings really took off. I was a young child when my parents bought their first record player. We all thought it was pretty amazing, but of course, vinyl records became insanely popular over the next 30 years. Various tape recording formats also came along at the same time, but they too were analog. All analog sound recordings suffered to some degree from noticeable hiss and distortion.

But that all changed abruptly in 1982 - Sony and Philips introduced the compact disc. It sampled the sound 44,000 times per second, and recorded it as a stream of 16 bit digital numbers.

The switch to digital music wasn't instantaneous. But over the next 15 years, CD recordings pretty much totally replaced vinyl records. Vinyl records wore out after being played a couple of hundred times. Tape recordings also wore out with time and were plagued by background hiss. CDs were perfect in comparison. The sound was superb, free of hiss and distortion, and CDs never wore out. Digital music has evolved, moving to iPods and then our smartphones. So CDs like vinyl records are mostly obsolete today, but music and audio have gone permanently to digital!

The transition to digital television took a little longer. It didn't happen until the late 1990s. That's because processing digital video requires very high speed processors and very complex integrated circuits. Even though every PC, every smartphone, and every tablet today handle video with ease, the technology either didn't exist or was way too expensive for commercial television to go digital before the late 1990s.

The benefits for television switching to digital are huge and similar to the benefits of digital audio. Digital TV results in a huge improvement in picture quality. Analog was plagued by all kinds of distortion and interference which all disappear when TV is broadcast digitally. For example, snow, grainy pictures, and ghost images go away with digital - the picture is either there or it's not. Color distortion also disappears. Compared with analog, the digital picture is always perfect.

High definition TV is not the same as digital. You can do high definition with analog, as Japan did in the early 1990's. And you could have digital standard definition as well. But in the U.S., we got them both at the same time. Digital high definition TV sets became available in 1998, and cable and satellite broadcasters both started supplying digital high definition content in 2002.

High definition was another huge improvement in picture quality. Standard definition TV has 525 lines. The analog signal has a resolution about equal to 700 pixels across. Some of this potential resolution actually extends off the screen a little, so standard definition TV is generally regarded to be equivalent to 640 pixels wide by 480 pixels high. In comparison, HDTV is 1920 pixels wide and 1080 pixels high. So HDTV has 6.75 times more pixels and detail than standard definition!

Digital television isn't only about how the video is transmitted. The television set in your house is also completely digital. TVs used to use picture tubes that were essentially analog devices. Electron beams for each color would light up the screen, but keeping the colors aligned with each other was difficult. They would become misaligned, distorting the picture, and requiring frequent repairs and maintenance. Today's flat panel displays are fully digital and don't suffer from these kinds of problems. Again the conversion to digital has eliminated a bunch of problems and made the picture perfect compared to analog.

Of course, we can't forget about digital photography! Digital cameras would seem to be much easier technology than digital television, but they took a long time to actually come to fruition. That's not just because of technology, but also cost. Film cameras in the 1980's and 1990s were relatively inexpensive and took beautiful pictures. Digital cameras couldn't replace them until the cost of high quality digital cameras could be brought down to the cost of film cameras. That didn't really happen until we were into the 2000s.

The history of digital cameras though starts in the early 1970s, just like other digital technologies. The first major development was the CCD or charge couple device.

Charge Couple Device IC

When we talk about how many mega-pixels a camera has, we are talking about how many tiny little regions on this device can store an electric charge. The electric charge these devices store is directly proportional to the light that has hit them, as they convert photons of light into electric charges.

Today, CCDs with 10-20 megapixels are commonplace. You probably have one in your smartphone. But the first CCD, which was introduced by Fairchild Semiconductor in 1973, was only 100x100 pixels. That's 10,000 pixels or 0.01 megapixels! Not much to work with.

The camera generally recognized as the first digital camera was a prototype developed by Steven Sasson in 1975 at Eastman Kodak. He built it with a Kodak movie-camera lens and the new 100 x 100 CCD.

The camera was the size of a toaster and weighed about 9 lbs. It took 23 seconds to store a picture.

The first digital camera

During the 1970's and 80s, there were some experimental filmless cameras built. I call them filmless, because they were electronic cameras, but not really digital. For example, they might start with an analog TV camera and store a picture from it on magnetic tape.

Real digital cameras didn't start to appear until the 1990s, and as we said earlier, they were very expensive and couldn't compete with the picture quality of film.

PowerShot 600

An example is the Canon PowerShot 600 that came out in 1996. It had a 1/3-inch, 832x608-pixel CCD, built-in flash, auto white balance and an optical viewfinder as well as an LCD display. It was the first consumer digital camera able to write images to a hard disk drive, and it could store up to 176MB. It cost $949. But the picture quality wasn't that great. Its CCD was less than 1 megapixel, so it wasn't good for anything other than small snapshots.

But progress never stopped and digital cameras have just gotten better and better. Once we got 10 megapixel CCDs, digital imaging became every bit as good as film, and the picture quality is today limited by the quality of the camera lens, not by the number of pixels. Another major development is flash memory cards that have come along in the last 20 years. Today we can store thousands of high resolution pictures on a single flash memory chip. Today even inexpensive digital cameras come with an immense array of features and capabilities that weren't available on any camera at any price just a few years ago. And most digital cameras today are also capable of taking HD video. Like everything else digital, it's the benefits of being able to put millions of transistor elements on a small single chip of silicon that makes it all possible.

There is still another product that began its development in the early 1970s, but took a while to become a common consumer item – GPS or Global Positioning System. At first it was a U.S. military program only. The military launched its first GPS satellites in 1974 and had a first generation system working by 1985.

GPS allows users to pinpoint their location by measuring their distance from at least four GPS satellites. This is accomplished using a lot of digital technology in the receiver and atomic clocks on the satellites. The satellites can transmit their exact location and a very accurate time. The receiver can compute its exact distance from the satellite by measuring the time the signal took to reach it at the speed of light.

From 1989 - 1993, the military launched a new set of 24 satellites that became the modern GPS we have today. During this same period, there was a big push to make the GPS system available to everyone, particularly commercial aircraft. At first, there was a low accuracy version made available to the public, but in 2000, the high accuracy version became available to everyone.

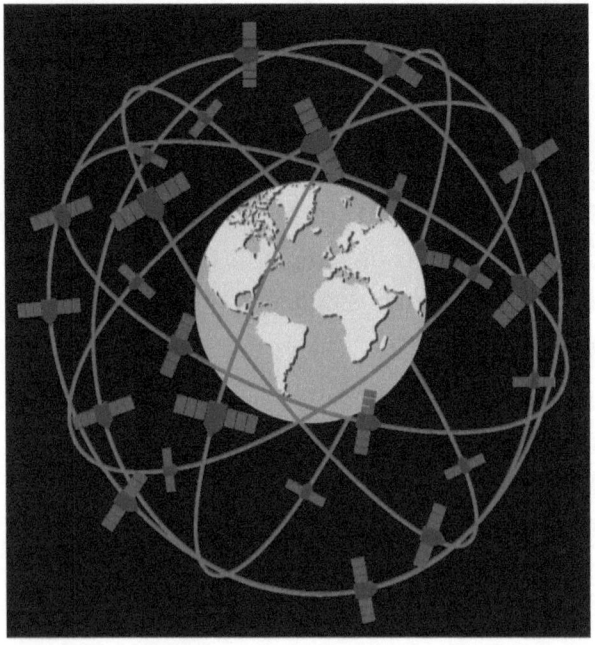

The first hand held GPS navigation device was built by Magellan Corporation in 1989. The first GPS enabled cell phone was introduced in Europe by Benefon in 1999. Then in 2001, in-car navigation systems from Garmin and TomTom began to appear. Finally in 2004, Qualcomm announced "assisted GPS" technology, allowing smartphones to use their cellular signal in combination with GPS signals to locate the user to within feet of their actual location.

In closing this chapter, we should reiterate that all of the products and technologies we've talked about in this chapter came about as a result of large integrated circuits that first came into existence in the early 1970s. And we should also mention that, thanks to the really large integrated circuits that exist today, every one of the products mentioned in this chapter are built into your smartphone!

Laptops, Tablets and Smart Phones

An amazing transition has occurred since 2000. At that time, personal computers were mostly desktops. Today, smart phones have totally taken over. Smart phone use totally dwarfs the use of desktop computers!

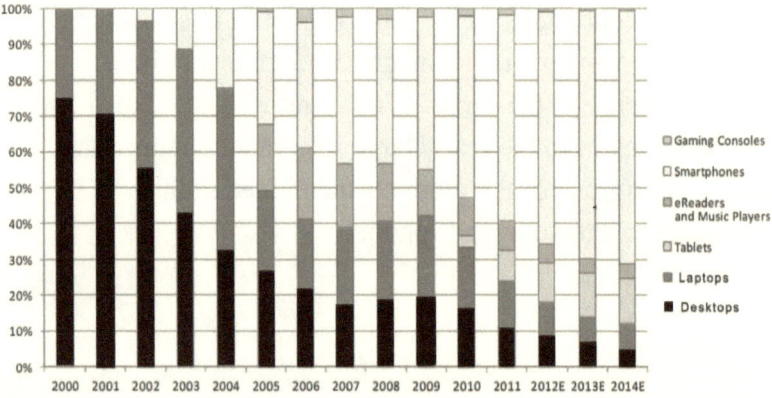

It's not just smart phones vs desktops either. Sales of laptops exceed that of desktop computers, and tablets and e-readers exceed the sales of desktop computers as well.

You might think that the trend you see in the chart above must have started in about 2000, where the chart starts. But actually, this evolution toward powerful portable and handheld computers started many years earlier.

In our discussion of the origins of the Apple Macintosh in the early 1980s, we discussed Alan Kay who worked at Xerox PARC in Silicon Valley. Kay was a true visionary about where computer technologies were headed. He was working on graphical user interfaces at PARC before anyone had them. He was one of the inventors of object-based computer languages and object-oriented programming, which are in widespread use today.

Alan Kay

Starting in 1968, Alan began developing a concept he called the Dynabook. It was a thin, portable, battery operated, personal computer. It would have enough permanent memory to store 500 pages of text or a few hours of audio. It would cost no more than $500. It would have a graphical user interface and would be, as much as anything, a learning tool for children.

The Dynabook doesn't sound too revolutionary today, because it is a rough description of every tablet, laptop, and cellphone in existence. But in 1968, it was a completely futuristic idea. Personal computers weren't in existence yet. A computer that came anywhere close to the capabilities that Alan Kay described would fill a small room, and consume enormous amounts of electricity.

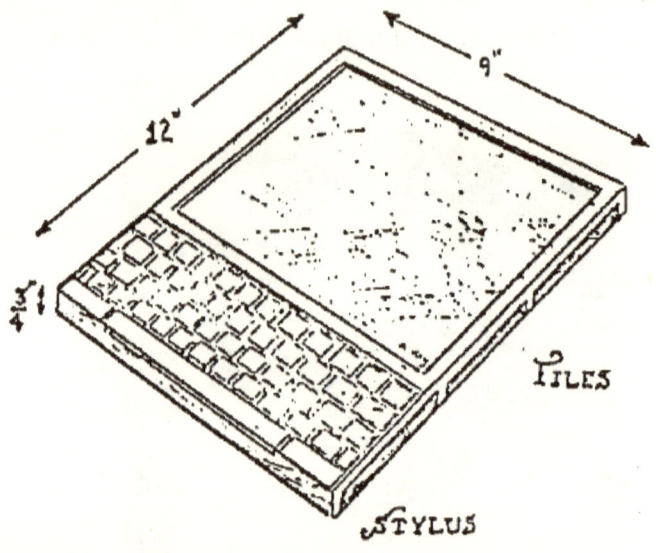

A sketch of the Dynabook from an article Kay wrote in 1972

Steve Jobs knew Alan Kay, knew of the Dynabook concept, and used it as a vision for the direction of Apple computer. It started with the Mac itself, and you see it again and again in the directions Jobs took Apple. The Mac Portable, the iPad, and the iPhone can all be thought of as Steve Jobs' implementations of Alan Kay's Dynabook vision.

But Alan Kay in 1970 was about 50 years ahead of the computer technology necessary to make his Dynabook vision a reality. In 1970, we didn't have processors that were fast enough or energy efficient enough; we didn't have lithium ion batteries; we didn't have low cost hard drives, let along the solid state hard drives that were really needed; we didn't have high resolution, flat panel full color displays; we didn't have the Internet or any other way to get the information that people would want on the Dynabook. So it's really only now, about 50 years later, that we are actually seeing the Dynabook become a reality.

The road from Dynabook concept to the modern laptop computer was not a smooth path. The first portable computers left a lot to be desired, and it is somewhat subjective as to when they might start to be called laptops. I will just briefly show you a few models that illustrate the progress.

The Osborne 1 above came out in 1981 and is generally regarded as the first portable computer. It used a Zilog Z-80 processor, so it wasn't an IBM compatible, but it came with a word processor and spreadsheet program for $1895. It was not battery powered; it needed to be plugged into a wall outlet.

Another early entry was the Compaq Portable Computer. It came out in 1982 and was 100% IBM compatible. It weighed 28 lbs. and cost $3000.

The Mac Portable came out in 1989. It was battery powered, and sold for $7300 with a built in hard drive. It had all the features of a MAC, but it weighed 16 lbs., mainly due to its lead-acid battery. Its price and limited battery life kept sales low.

The IBM ThinkPad came out in the mid 1990's. It's a good example of what we now consider a laptop. The Model 770 from 1997 is shown above.

Tablet Computers had a rough start much like laptops. It took a while for the necessary technologies to mature, and some of the early entries struggled with high prices and poor market receptions.

The GRiDPad shown above was probably the first tablet. (The capitalization in GRiDPad is theirs, not mine.) It was introduced in 1989 by Grid Systems Corp. and sold for a whopping $2370. Like most tablets, the GRiDPad used a stylus and touch sensitive screen. It ran MS_DOS and had limited success as a result of its high price.

Another early entry was Apple's Newton shown above. It was introduced in 1993 and was priced at $700. It was the first device to coin the term PDA for personal digital assistant. The Newton was also the first tablet to attempt handwriting recognition. It was an interesting novelty, but made a lot of errors. (I can say from my own personal experience with a Newton that the handwriting recognition errors were very annoying.) The Newton was also a little too big and heavy to live up to expectations, so its success was limited.

In 2000, Microsoft introduced its first tablet. Bill Gates expected tablets to quickly replace conventional computers.

He guessed wrong about how fast they would catch on, but more importantly, he targeted the Microsoft Tablet PC at business and made it a full blown computer running Windows 95 (and later Windows XP). It failed to catch on with both business and consumers, and Microsoft really didn't succeed in tablets until years later with the Microsoft Surface.

Amazon's Kindle shown above was first introduced in 2007. As it's primarily a reader, it is of course a special kind of tablet, but probably qualifies as the first really, really successful tablet. It has gone through a number of iterations and remains popular today.

Another successful tablet is Apple's iPad. First introduced in 2010, it too has gone through repeated upgrades over the last several years. It's probably the best example of Steve Jobs' quest to make Alan Kay Dynabook a reality. It remains the most popular tablet.

Finally, Microsoft's Surface is another successful tablet. It is partly tablet and partly laptop, but is very popular, and competes well with Apple's iPad.

But the real star in today's computer technology is the smartphone. As we said at the beginning of this chapter, smartphones today amount to 70% of all personal computers on the planet. You might want to question whether smartphones should be classified as computers at all. But they are very sophisticated computers, and they come along with a whole bunch of accessories. I am going to take a quick side trip and talk about all the technology in a modern smart phone. But it will be a very quick tour, because there is so much technology crammed into your smart phone that we could write a whole book just on the inner workings of a smart phone.

We all have smart phones today. We all know how to use them. And we all have a pretty good idea of what they will and won't do. But the modern smartphone is an incredible array of technology. The electronics in a cellphone would have filled several rooms just a few years ago. It contains thousands of subsystems, millions of lines of computer code, and the collective work of tens of thousands of people. It is easy to say that no one individual understands more than $1/1000^{th}$ of the collective technology in a cell phone.

To give you some idea of a cell phone's complexity, let's just look at some of the pieces, each of which is a mysterious black box even to most of the people who work in the cell phone industry.

There is an audio system, where sound from the microphone goes in, and sound received is sent to the speaker. The sound is digitized coming in from the microphone and it is converted back to analog audio on its way out to the speaker.

There is a radio frequency or RF section that maintains two-way communication with nearby cell-towers. It not only communicates digitized voice in both directions, but also communicates with the cell-tower, identifying itself and associating your phone with your account, and facilitating the handoff between cell-towers as you move about.

There are additional RF sections that handle data communication between your phone and the cell tower, and also with your local Wi-Fi network. There's also Bluetooth communication with earbuds and other accessories.

There is a full-blown computer system - a very sophisticated one with all kinds of sub-systems that maintain your pictures, your apps, your email, Internet access, text messaging, GPS navigation, your contacts, reading material, utilities, maps, etc.

There is a camera system – itself with numerous sub-systems.

There is a system that stores and plays all your music. We could go on. It would take a whole book to go through the basics of how all this stuff works!

We've talked about the early market entries and the successes in laptops and tablets. Now let's do the same thing with smartphones.

A smart phone by definition is a cell phone which can perform many of the functions of a computer, typically having a touchscreen interface with the ability to text, access email, access the Internet, and run downloaded applications. The first smart phone to meet these criteria was IBM's Simon, introduced in August 1994. It was way ahead of its time. It could send email and faxes, had a calendar, world time clock, and an address book. It didn't have Internet access because it was too early. The World Wide Web had just barely been invented.

Simon was way ahead of its time, but it wasn't a big success. It was way too big (8 inches long), way too expensive ($1100), and had very poor battery life (about an hour's use per charge). Simon sold about 50,000 units and only stayed on the market a few months.

There were many smart phones introduced in the late 1990s. The Nokia 9110i Communicator introduced in 1999 is a good example. It did everything IBM's Simon did, plus had a keyboard and Internet access. It was smaller (6 inches long) and had a lithium ion battery that could last about 6 hours.

Blackberry 5810

Blackberry was another very successful early smart phone. Research in Motion or RIM, a Canadian firm, introduced the first Blackberry in 1999. It was a pager, not a smart phone, but already had many of the features we associate with a smart phone. In 2002, they introduced the Blackberry 5810, which was a smart phone. Blackberry had a large, very loyal customer base for many years. Its strengths were its popular email system and the Blackberry proprietary operating system.

Then in 2007, Apple launched the iPhone. It was the first of modern smart phones to have many of the features we take for granted today:

- A screen you can navigate and zoom in and out with your fingers.

- A built-in camera with video

- Wi-Fi capability

- GPS and Google Maps integration

- A battery that can be expected to last all day through a variety of uses

- A small size and weight that makes it a truly hand-held computer

- A high resolution color display

- Full feature web browsing

With all the new features that came with the iPhone, the nature of the smartphone market changed very quickly. Up until the iPhone, smart phones were mostly seen as business tools – not consumer products. The iPhone completely change that! All of a sudden, everyone on the planet had to have a smartphone!

The iPhone definitely created a new standard for what we expected from a smartphone. At the time of its introduction, Apple said it was 5 years ahead of all other smartphones. They might have been true, but the gap closed very quickly.

Quietly, without all the fanfare that had accompanied the iPhone, the Android operating system was under development in 2007. It was an open-source project backed by Google. It was planned as a first-class smartphone operating system that could be used by any smartphone maker, as long as they used Google's web browser, search engine, and maps. It obviously needed some additional work when the iPhone appeared on the scene, but the race was immediately on to make Android competitive with the iPhone.

In May 2008, the Chinese company HTC (High Tech Computer Corp) announced it would be building smartphones using the Android operating system. Motorola, Samsung and a host of other smartphone manufacturers soon followed. Over the next couple of years, the Android operating system matched most features of the iPhone.

Today, Android phones and iPhones are very similar products. The iPhone still has an enormous following of dedicated fans that wouldn't have anything else. But Android phones have dedicated fans too! There is one important difference between iPhones and Android phones. Android phones are much less expensive, due in large part to multiple suppliers and increased competition. As a result, iPhones have about a 20% share of unit sales of smart phones worldwide, while Android suppliers share most of the other 80%. In dollar sales of smart phones, iPhone has about 40% of sales worldwide.

Motorola Part III

By the mid-90s, I had spent several years working in Motorola's Analog Integrated Circuits Division. With everything going digital, you might think that the Analog Division was the wrong place to be, but our business was doing very well. Analog devices are necessary to interface digital electronics with the real world. So cellphones contained several of our parts. Power supplies and battery chargers were also full of our components. So Analog ICs was a good business.

At my desk at Motorola Analog IC Division in about 1990

Motorola Semiconductor was truly an international company, and I got to see much of the world traveling on business for Motorola. We had factories and/or engineering facilities in Mexico, Japan, China, Korea, Hong Kong, the Philippines, Malaysia, France, and the Czech Republic. And that was just for the products I was involved with. There were many other international locations if you looked at the Motorola's entire semiconductor business.

Motorola had always been a great place to work, considered one of the country's premier employers. It was quite common to spend your entire working career with Motorola, and I was one of the people who had done just that. The semiconductor industry had always been famous for boom and bust cycles, so we had layoffs from time to time, but most people felt pretty secure about their jobs after they had been there for a while. And Motorola had always been a good company to work for financially. They had good medical coverage, a profit sharing plan, stock options, and continuing benefits after retirement.

I am telling you all this to set the stage for what happened next. As we stated in a previous chapter, Motorola began to lose market share in cell phones in the mid- 90s. It made a huge strategic error by misjudging how quickly digital cell phone technology would replace analog, and found itself way behind the competition. Suddenly, the entire focus of the company was staying successful in cellphones. Everything else, including its semiconductor business, suffered while all the companies resources were poured into its cellphone business. Not only that, but Motorola Semiconductor's largest customer was Motorola's other businesses. So as Motorola Cellular lost market share, so did Motorola Semiconductor. In the 1980's, Intel, Texas Instruments, and Motorola competed to be #1 in semiconductors. By the end of the 90's, Motorola Semiconductor had slipped to about 10th place.

Motorola began selling off pieces of the business. It needed to focus on its core business which by then was cellphones. And it needed cash to continue investing in cell phone technology and production.

One of the things Motorola wanted to sell off was parts of its semiconductor business. We reorganized in 1998 into the pieces Motorola wanted to keep, and the pieces that it wanted to sell. They wanted to sell the discrete components like transistors and rectifiers, standard logic ICs, and about 1/3 of their Analog IC products. So a new organization was formed, called the Semiconductor Components Group, consisting of the pieces they wanted to sell.

In August 1999, Motorola sold its Semiconductor Components Group to an investor group and it became ON Semiconductor. I went to ON Semiconductor, but then retired in 2001.

During my last couple of years at Motorola, it was a very difficult environment to work in. Reorganization was constant and perpetually stressful. Lay-offs and early retirements became more and more common. Motorola was no longer the proud company it once was. Stock options that we once thought were a big component of our retirement benefits became worthless. For all of us who had devoted our entire career to working at Motorola, it was very disheartening. We had to watch the company we had been so proud of and worked so hard to build gradually disintegrate before our eyes!

So my relationship with Motorola ended there. Then in December of 2004, Motorola sold off the rest of its semiconductor business, which became Freescale

Semiconductor. But what happened to Motorola since then? And how has it faired in the era of smart phones?

In 2004, Motorola introduced the RAZR smart phone. It was one of the best-selling and long lasting cell phones on the market. But as the smart phone business continued to evolve and mature, Motorola fell further behind

Motorola RAZR

Motorola relied for too long on the RAZR and was late in offering a new phone with G3 and new touchscreen features. Samsung and LG both passed Motorola in market share, just as Nokia had previously done.

By then, Motorola Inc. had broken up into several smaller companies. The cell-phone portion was called Motorola Mobility. In August, 2011, Google announced it was acquiring Motorola Mobility. It did so, for the most part, to acquire Motorola's patent portfolio and give Android phones a strong patent position.

Then, less than three years later, in January 2014, Google announced it was selling Motorola Mobility to Lenovo, the same Chinese company that acquired the PC business from IBM. Lenovo continues today to make Motorola cell phones, but has dropped the Motorola brand name in favor of Moto.

Moore's Law Revisited

Way back in the beginning of this book, we talked about Gordon Moore's Law. He said that semiconductor technology would progress so rapidly that we could expect the number of transistor elements on a single IC chip to double every two years. It turned out he was pretty conservative, because it's doubled every 18 months for about 50 years now.

This technology trend has brought us better and better computers, digital cameras and televisions, smartphones and tablets, the Internet, and changed our lives significantly over the past 50 years. Let's take a minute to return to the semiconductor technology that brought us all this, and take a look at what Moore's Law actually means with some pictures and a few comments.

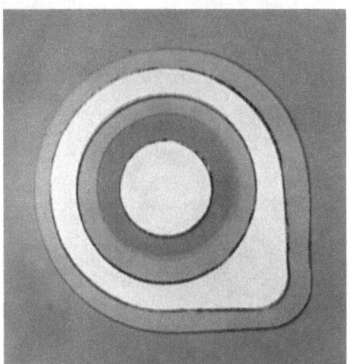

This is the top view of a single transistor chip from the early 1960s. Like everything else we'll look at, it's made of silicon. The chip might be 10 x 10 millimeters (3/8th in. on a side) and the size of the smallest feature is roughly 1 mm or 1000 microns.

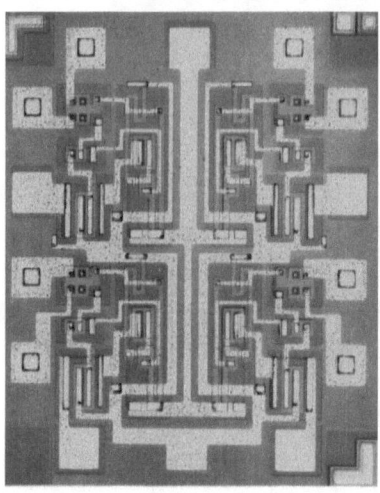

This is a simple integrated circuit from the mid-1960s. Specifically, this IC contains 4 NAND gates. A NAND gate is a fairly simple logic element. The chip has 16 transistors and its smallest features are about 50 microns in size.

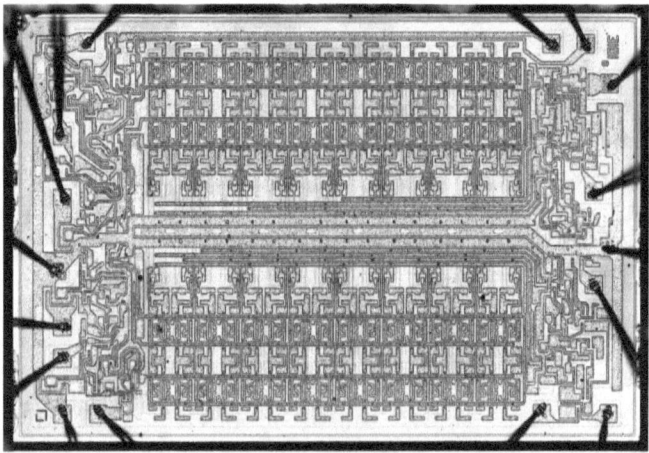

This is Intel 3101, a 64 bit static RAM chip from 1969. It contains approximately 500 transistor elements and its smallest features are 8 microns in size.

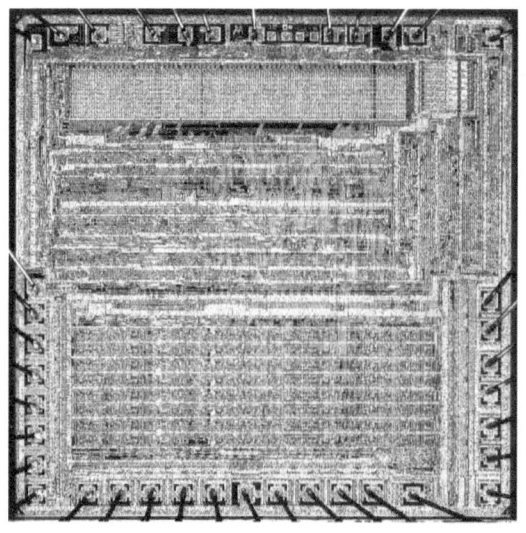

This is the 6800 microprocessor from 1974. It has 6800 transistor elements and its smallest features are 6 microns in size.

This is the Pentium microprocessor chip from the mid 90's. It contains 3.1 million transistor elements and its smallest features are 0.35 microns in size.

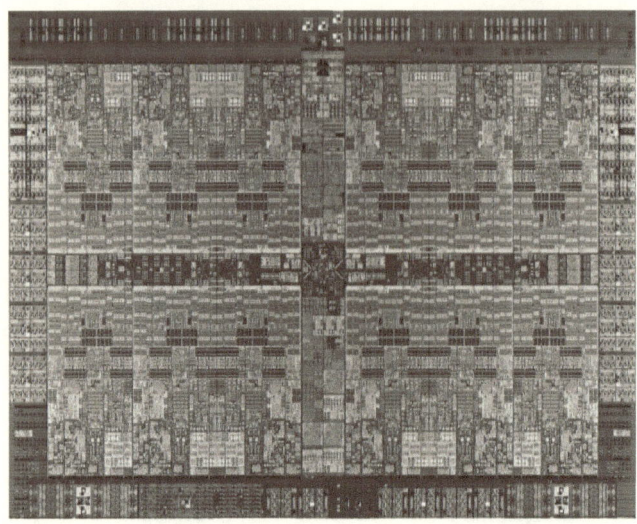

This is an IBM Power8 processor from 2014. It contains 5 billion transistor elements and its smallest feature size is .022 microns or 22 nanometers.

So far, we have talked about Moore's Law as though he proposed it and then it just magically happened. In reality, nothing could be farther from the truth. The continued shrinking of transistors and ability to put more and more functionality on a single chip of silicon was the result of hundreds of thousands of brilliant engineers and scientists working in a number of different fields. We will now look briefly at some of this work.

Improved Photolithography

Integrated circuits are created on a wafer of silicon using a series of photographic images, shining light through a mask to expose a photosensitive resin, which is then etched. The smallest features that could be resolved by this process went from about 100 microns in the early-1960s to about 5 microns by the mid-70s. Today we are down to about 10 nm. That's a 10,000 times improvement in 60 years.

What made this possible was mostly improvement in the equipment. The masks for the different process steps have to be aligned with each other on the silicon, so the tool that exposes the silicon wafer through the mask is called an alignment tool. In the beginning, a whole wafer was exposed at once. Getting all the masks to align across the wafer was plagued by many issues. Full automation of the alignment process helped. But as circuits got smaller, just temperature variations caused misalignments. So alignment tools were replaced by steppers, which only exposed one die at a time. This greatly improved our ability to put small features on silicon.

But as feature size continued to decrease, the wavelength of the light used to expose the wafer started to become a problem. The solution was to move to shorter wavelength light, first to violet, then ultra-violet, and then deep ultra-violet. As this process continues, we will eventually be forced to use x-rays, which has a whole new set of obstacles to overcome.

Optical Shrinks

One of the more straightforward ways to make an integrated circuit smaller is called an optical shrink. To take advantage of the previously discussed improvement in photolithography, you just reduce the size of the whole chip, making everything half the size it was before. You don't change anything else; just photographically do a one half reduction. Since the chip is two dimensional, you get four times the number of chips on the same wafer!

For a number of reasons, you can't just keep making optical shrinks forever. Eventually you have to redesign the entire chip and do a new layout to continue shrinking the die. However, it is sometimes possible to do more than one optical shrink as process technology improves.

Smaller, Simpler Devices

Another major way integrated circuits have continued to shrink over time has been the evolution of transistor design. We started out building ICs using bipolar transistors, but they were big and clunky compared to field effect transistors (FETs). FETs were simpler than bipolar transistors and simpler means smaller. FETs with metal gates (MOSFETs) were simpler and smaller yet. Almost all integrated circuits are done with MOS technology today. This is but one example of how individual components themselves within an IC have gotten smaller over time.

Ion Implantation and Metal Deposition Sputtering

As the dimensions of everything on the chip continued to shrink, other parts of the process had to adapt to keep pace. Diffusion furnaces were used to drive impurities into silicon to form various parts of an IC. But as everything got smaller, the depth of these diffusions became more critical. For critical depths, diffusion furnaces were replaced by ion implanters, which could very precisely control depth. Metal layers used for metal gates and wiring up the circuit used to be evaporated onto the wafer. But to get more uniform deposition, sputtering system are now used. The shrinking size of ICs has been made possible by ever more precise and ever more expensive equipment at almost every step in the process.

Single Transistor Memory / Dynamic RAM

Memory is such an important part of IC technology that we will talk about it separately. There are of course stand-alone memory chips, but many ICs like microcontrollers have large amounts of memory built in. Static memory which holds onto its 1 or 0 as long as the power is on contains four transistors per bit. It's relatively big – what the industry needed was a way to get a single transistor to hold a 1 or 0, allowing 4 bits to be stored were one was before. That was accomplished in 1970 when Intel introduced the first dynamic memory chip.

Dynamic RAM only holds onto its 1 or 0 for a few milliseconds. At first that may make dynamic RAM sound worthless, but remember that microprocessors can perform many operations every microsecond. In this environment, a few milliseconds is a long time. Special hardware does what is called a refresh, where the memory's content is read and then rewritten every few milliseconds. This allows it to keep its content stored as long as the power is on.

All of the progress made as a result of the technology described above together means that Moore's Law is still true after 50 years. The progress we have made is mindboggling. Right now in 2019, some production fabs are producing IC using 14 nm technologies. Leading edge IC manufacturing will soon be in the 10-12 nanometer range.

There is, however, a limitation on the horizon. A silicon atom is only 0.2 nm across. Today's transistors are only 70 silicon atoms wide.

Moore's Law is still alive and well. And progress is expected to continue for the next several years. But at some point in about the next twenty years or so, we will need some significant breakthrough to allow this trend to continue.

If You're Interested:
Semiconductor Manufacturing

The incredible strides we have made in semiconductor technology and manufacturing is what makes today's digital electronics possible. So it seems reasonable to spend a little time talking about the actual manufacturing process and how semiconductors are made. As I started thinking about what should be included in this chapter, I realized it is an enormous subject – hard to summarize in a few pages. But I will try to give you an overview here, without getting into a lot of detail.

Let's begin with the facility where the integrated circuit chip is fabricated. In the industry it's actually called a fab. They process silicon wafers, which are slices of single crystal silicon. Each wafer contains anywhere from a few hundred to a few thousand integrated circuit chips.

A fab is designed to conform to a very specific process. For example, it might make CMOS devices with 20nm minimum feature size on 300 mm (12") wafers. If you want to make anything else, or want to make smaller features, or want to make them on 350mm wafers, you need to build a new fab!

A fab comes with a set of detailed specifications called design rules. Design rules spell out the capabilities of the fab. They tell a circuit designer what he can design and implement in this fab. They tell him the rules he must follow to successfully design a product that this fab can manufacture, and they tell him how it will perform electrically when it is produced by the fab. The design rules create a very important separation

225

between two processes: the process of designing the circuit and the process of fabricating it. It means that the circuit designer does not need to understand the inner workings of the fab, as long as he follows the design rules. And it means the engineers in the fab don't necessarily need to know what circuits they are making, as long as they conform to the design rules.

In reality, designing integrated circuits is heavily computer aided. Computer tools aid in laying out the circuit such that it conforms to the design rules, and other tools provide simulations of how the device will perform electrically. So the circuit designer has a lot of help. That's important because the incredibly complex circuits we have today would most likely be impossible without computer aided design.

Let's look now at the actual process. The fab starts with bare silicon wafers, but where do the wafers come from?

The picture above left shows a single crystal of silicon that is grown by pulling it from molten polycrystalline silicon. It is then sliced into wafers and the wafers are polished to a mirror-like finish. A tray of polished wafers is shown on the right.

At the right is a list of the process steps necessary to build a CMOS integrated circuit. The pictures are a side view of what is happening on the wafer.

To begin, silicon oxide is grown on the wafer and then openings are etched in the oxide. Through those opening, a diffusion furnace drives impurities into the silicon to form n-wells.

Then new oxide is grown and polysilicon is deposited. The polysilicon is etched away except over the gate regions. The oxide is removed in places and ion implanters add impurities to form the sources and drains for half of the field effect transistors.

New oxide is grown, etched in places, and ion implanters add sources and drains for the other half of the field effect transistors.

Then silicon nitride is grown over the wafer and etched away where metal will touch the silicon. Then the metal is deposited across the wafer and etched to form the electrical contacts to the device.

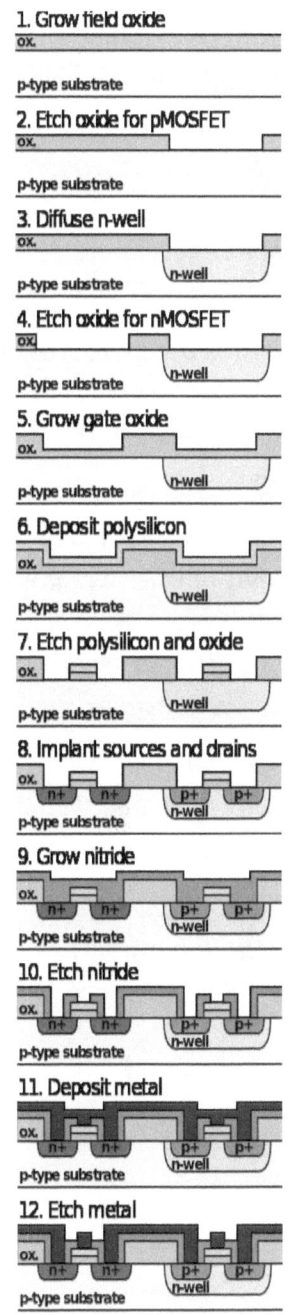

1. Grow field oxide
2. Etch oxide for pMOSFET
3. Diffuse n-well
4. Etch oxide for nMOSFET
5. Grow gate oxide
6. Deposit polysilicon
7. Etch polysilicon and oxide
8. Implant sources and drains
9. Grow nitride
10. Etch nitride
11. Deposit metal
12. Etch metal

But each time we say etch away oxide or metal or polysilicon from specific places on the wafer, we must go through a series of step to make that happen. This process is called photolithography.

A coating of light sensitive resin called photo-resist is applied across the whole wafer. A mask is aligned over the chip and ultra-violet light exposes the photo-resist through the mask.

The photo-resist is developed and the resist is removed wherever it was exposed to UV light. At this point, we can etch away the oxide, polysilicon or metal where the photo-resist has been removed.

And finally, we can remove the remaining photoresist, and the wafer is ready for the next step in the process.

The wafer must go through this photolithography process many times during the fabrication process.

a. Prepare wafer

oxide

substrate

b. Apply photoresist

PR

oxide

substrate

c. Align photomask

glass

Cr

PR

oxide

substrate

d. Expose to UV light

glass

Cr

PR

oxide

substrate

e. Develop and remove photoresist exposed to UV light

PR

oxide

substrate

f. Etch exposed oxide

PR

oxide

substrate

g. Remove remaining photoresist

oxide

substrate

Now let's take a look at some of the equipment in a fab.

Above is a diffusion or oxidation furnace. Some are used to grow oxide on wafers. Others are used to diffuse impurities like boron or phosphorus into the wafers.

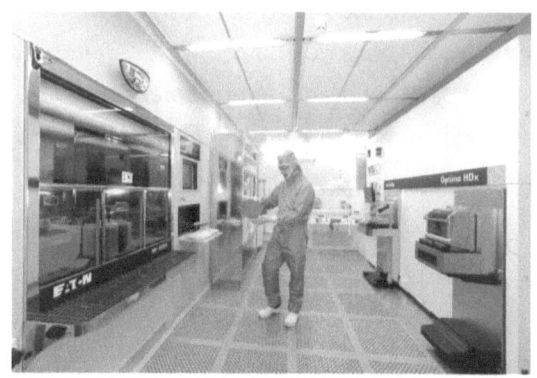

Above is an ion implanter. It is used to implant impurities in wafers when more precise control of those impurities is

required than can be achieved with a diffusion furnace. Notice the clothes the employee is wearing. Most areas in a fab are clean-room environments where the air is kept ultra clean and everyone inside must wear special clean-room clothing.

Above is an alignment tool, known as a stepper. It is used to align the mask and expose the photo-resist to UV light. It does this be progressively stepping across the wafer. These are typically the most expensive piece of equipment in a fab, costing 10s of millions of dollars each. There are many steppers in any fab. They are a big reason fabs are so expensive.

Above is a sputtering system. It is used to precisely deposit a metal layer on silicon wafers. This metal is used to connect the various circuit elements together and also to create the external contacts to the chip.

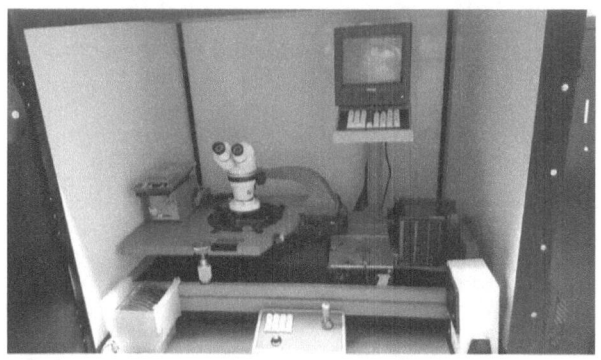

We haven't talked yet about testing devices. They are tested twice, once in wafer form as part of the fab process and again after being packaged. The test in wafer form is done on a piece of equipment called a prober as shown above. It sequentially tests each chip on the wafer and puts an ink dot on any device it finds defective.

*The prober is only half of the test setup.   The other half is the
actual test system, as shown above. It includes a computer and
set of power supplies and measurement instruments which can
put a part through a series of tests to insure it functions
correctly.*

The photo at the bottom of the last page is a wafer with the integrated circuits on it clearly visible.

Building a new fab is not cheap. A leading-edge fab today costs around $15B! The biggest semiconductor companies have many fabs located around the world. But because of the huge cost and the need to keep a fab running at full capacity, some companies elect to use subcontractors to fabricate their chips. They are understandably referred to as fabless semiconductor companies. Familiar names like Qualcomm, AMD, Broadcom, and NVidia are all fabless.

Now that we have talked about fabs, let's step back and look at semiconductor manufacturing in total. The main parts are as follows:

1. *Make wafers from single crystal silicon*
2. *Fabricate the circuits on wafers*
3. *Test circuits in wafer form*
4. *Saw up wafer into individual die*
5. *Mount the die on a lead frame and bond wires between die terminals and lead from pins*
6. *Injection mold plastic around the die*
7. *Punch finish part out of the lead frame and bend leads into position*
8. *Final test part to verify its functionality.*

We've already talked about the first three steps in the process. Now let's discuss the rest.

A singulation machine or die saw like the one above is used to cut a wafer into individual chips in preparation for assembly.

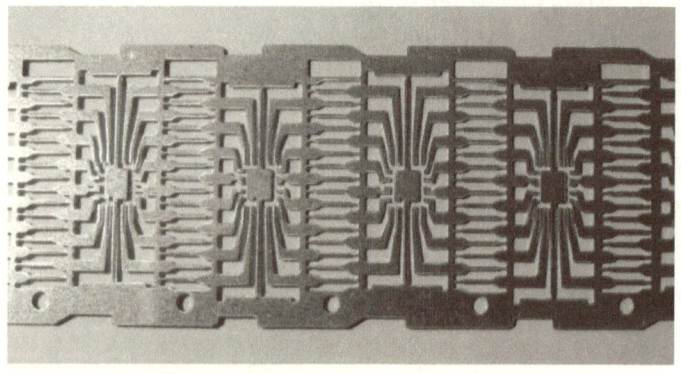

The chip or die is then mounted on a lead frame like the one shown above. This step is performed by an automated machine called a die bonder.

Another automated machine called a wire bonder attaches wires between the lead frame and the chip.

The lead frame and chip are then encapsulated by injection molding using a large mold press like the one above.

Then the molded lead frame goes through a trim and form process where individual devices are separated from the lead frame and the leads are bent into the proper position. Today all these steps are highly automated. And automated visual inspection systems are used to insure that all parts conform to physical specifications.

The device still needs to be tested electrically. An automated handler like the one above moves the parts to a test head, then tests, sorts and packages the good devices. As it was with the wafer prober, this handler is coupled with a test machine, and the two machines together comprise the complete test system.

Almost all assembly and test today is done in Asia. It can be done by the semiconductor company itself or through subcontractors. There are many very cost efficient assembly/test subcontractors throughout Asia, so it is very common for semiconductor companies to use subcontractors. Even companies with their own assembly/test facilities will most likely be using a combination of their own facilities and some subcontractors.

Semiconductor manufacturing technology is constantly changing. Wafers were 1" in diameter in the 1960s, and today they are 12" in diameter. There are plans to go to 18" in the next few years.

The smallest features on today's IC are 10-12 nm. We have had to go from visible light to deep ultraviolet light in photolithography to get features that small, because the wavelength of regular light is longer than the features we are trying to resolve. So there are plans to go to even shorter wavelength light or even x-ray lithography to continue making smaller and smaller transistors.

There are also some ideas for super-automated fabs, where everything is totally automated and human intervention is completely absent.

Fabs are huge facilities today, like this Intel fab in Arizona. In the early days of semiconductors, a building like this might have housed four or five fabs, as well as an assemble/test site.

Observations about Today's Computers and Their Operating Systems

One of the most amazing things about computers is the evolution of hardware over time:

	Original PC	Today's PC or MAC	Ratio
Year	1981	2019	
Processor	Intel 8088	Intel Quad Core	
Processor Speed	4.77 MHz	3 GHz	630
Data Width	16 bits	64 bits	4
RAM	64 KB	8 GB	125,000
Mass Storage	160 K - 5 ¼" floppy disk	1TB Hard Drive	6,250,000
Typical Display (pixels)	320 x 200	1920 x 1080	32

It has only been 35 years since the original IBM PC.   Yet there is a 630 times improvement in processor clock rate. Combine that with a 4x improvement in data path width and four cores, and you get a 10,080 times performance improvement. Available RAM has increased 125,000 times. Mass storage has increased 6.25 million times. The Internet has made the whole thing 100 times more useful. These trends are continuing, and where they are going to evolve is a very good question. But another big question remains: how have users benefited from all this?

In my opinion, we have very little to show for this incredible increase in hardware performance and storage space. I can remember around 2000, when personal computers finally got fast enough to process digital video. Before that, digital video required very expensive specialized hardware. So that was a major benefit from this increased performance. And around 2000, also, we got the Internet, which very quickly made personal computers 100 times more useful. Games got better, faster and more realistic. We can open and close programs faster. But for the most part, the user experience has not improved as much as one might think given these astronomical improvements in hardware performance.

Huge, bloated, multitasking operating systems are a big reason much of the hardware improvements haven't changed the user experience.

A good example of this is the amount of time the operating system consumes trying to keep viruses, hackers, and other invaders from commandeering your system. This might not be necessary, if the operating systems were built right in the first place. Just like the Internet, these operating systems were designed and conceived before we knew security would be such a problem. They have never been redesigned from scratch to make them secure. Instead, we have an endless list of modifications tacked on to an unsafe system, in an effort to try to protect it.

Operating systems used to take up a few hundred thousand bytes of disk space and RAM. Today they take up a gigabyte of RAM and several gigabytes of disk space. You can rightfully argue that today's operating systems do 100 times more than they did 35 years ago, but why are they 2 million times bigger?

Similar things can be said about applications. Today's applications are typically gigantic collections of programs and files, with everything, including the most time consuming operations, written in very high level languages, with almost no effort made to optimize execution and performance, except in the areas of image and video processing. These applications are distributed among hundreds or sometimes tens of thousands of different files that have to be accessed as the application is running. In their efforts to constantly add new features and introduce new versions, application developers have created slow, lumbering giants that don't come close to actually taking advantage of the enormous amount of computer power in today's typical personal computer.

So now let's pose some questions about why modern operating systems and applications do some of the things they do. And what might we do differently?

Back in 1977, when I was writing my own disk operating system, I found that, traditionally, hard drives were divided into sectors; sectors were grouped into larger units called clusters. Then clusters were dynamically allocated by the disk operating system, with a bunch of tables used to control how files were distributed over this very expensive commodity called disk space. A typical file could end up spread around on the disk, and as it was modified and perhaps grew over time, it became more and more fragmented.

For my operating system, I decided to keep things simple. Files got laid down in sequential sectors until the whole file was stored. It could be changed and restored in the same place, but if it grew, and was now too big to fit where it did before, the space it previously occupied was marked as unused, and the file was saved in a new location. This system's advantage was not only its simplicity, but speed as well. However, it caused the disk to fill up too quickly, leaving behind lots of unused and unusable space. The fix was pretty simple – I wrote a utility program that pulled everything forward, filling all the unused space, and freeing up space at the end that was then available for use again. It just needed to be run once in a while when the disk got 70 or 80% full. It worked great. Applications loaded instantaneously.

As hard drives have grown in size and become very inexpensive today, I believe my simpler and much faster way of doing things makes more sense than ever. But as far as I know, we are still doing things the same old way that IBM started doing it on mainframe computers back in the 1960s.

Today, even a simple application contains hundreds of separate files. If I were writing a disk operating system today, I would need to do things a little differently than I did back in 1977. Perhaps we would need a table that tells the system about blocks of files that need to be loaded together, but we would still store them in order and load them all at once. It would be tremendously faster than rummaging around all over the hard drive just to load a simple application.

Now let's change subjects. Why does the operating system allow every application to have stuff running in the background on your computer, even when the application itself isn't running? The application designers themselves like this, because they can do stuff like interrupt you to tell you they have an update available. Or they can open the application automatically when you do something that requires the application – example: iTunes opens up when you connect your iPod. But all this stuff adds a lot of overhead, slowing down your operating system and the overall performance of your computer. Eventually performance degrades to the point you need a new computer. Perhaps this is the answer to my original question – why is this allowed?

Even if the operating system allows applications to run stuff in background, shouldn't it at least offer us some control over it. For example, I use my web browser, my email program, the operating system itself, and a few other applications every day. I don't mind that they may have to run some tasks in the background. But I also have dozens of applications I only use once or twice a year. I certainly don't want them running anything in background. I don't want them slowing down my computer. I don't want them rummaging around on my hard drive. I don't want them looking for updates. I don't want them doing anything except at the one or two times a year that I actually need them. This is pretty common sense kind of stuff that someone writing an operating system should have thought about a long time ago.

Another thing I have never understood about modern operating systems is why we completely rebuild the operating system every time you power up your computer. It takes at least a minute and sometimes as long as five minutes for the computer to be ready to use. Why not just take an image of the computer's RAM, store it on the hard drive, and, assuming there was nothing wrong with the operating system when you shut your computer down, just restore it to the exact same configuration as when you shut it down. Again, I did it in 1977 - it worked great, and the computer booted up almost instantly.

Personal computers have been around long enough now that they should be appliances. They should be reliable. A basic version of the operating system should be in Flash memory, so that no matter what happens to an application or the operating system, you can always instantly get back to basic functionality by pushing a single button. And we as users shouldn't have to worry about viruses and security issues.

Unfortunately, these things aren't going to be realized by just another operating system update. Real improvements require a complete, clean rewrite of the operating system, simplifying it, streamlining it, and making it inherently secure. It's not obvious that is going to happen any time soon.

Cybersecurity

You might be surprised to see a chapter on cybersecurity in a book about the history of digital electronics. But over the past 20 years, the security of the Internet has become a major topic of discussion for every consumer, every business owner, anyone who does online banking or online purchases. So we will look at it here – how has it gotten to be such a problem? And what is it going to take to fix it?

What a nuisance all this has become! Everything needs a user name and password. But don't use the same password over and over, because some system is always getting hacked. If the hacker gets into one of your online accounts, you sure don't want them to get into all of them. And don't use anything for a password that you might remember, like your pets name and your favorite number. Your password has to be random. It has to be at least 10 characters long. It has to contain capital letters, numbers, and at least one symbol. And when you're done entering your password, you need to type in the series of distorted characters in the CAPCHA box - we need to make sure you're not a robot trying to get into people's accounts. OK, now just to be sure, answer one of those personal questions so we can be sure it's you. Oh no! You must have done something wrong! We will need to send you a text message with a 6 digit code, so you can prove you're you.

This goes on all day, every day for every one of us. And yet, hackers are constantly getting into our personal information or worse yet, compromising our credit cards and bank accounts. And it is still getting worse. You might think that, with all this added security, hackers would be finding it more and more difficult to get into these systems. After all, governments and business all have cybersecurity staffs whose entire job is to keep your data safe and hackers out of our systems.

But as we all know, the problem continues to get worse. Cybercrime is very profitable. I don't know how many hundreds of thousands of people there are out there working full time to protect our data and money, but I'll bet there are 10 times more than that trying to steal our data and money!

The problem is a flawed structural design. The Internet and networks in general were designed to be simple, easy to use tools to move information around. No one considered online stores, online banking, and online investing. No one ever thought we would be storing classified information on the Internet. After all, the Internet started out as a way for universities to exchange knowledge and read each other's papers.

The Internet was never designed with security as its most sacred and primary objective. Security has come as an afterthought with an endless list of mostly unsuccessful patches in an attempt to fix and make it secure. But it was never secure to start with. And it has gotten much, much worse over the years, because we have allowed a few big companies to amass huge amounts of priceless personal data in huge datacenters and server farms. These data centers are like Fort Knox – not because they are hard to break into, but because no matter how well they are protected, what's inside is priceless and very tempting. So endless numbers of hackers are always trying to exploit some undetected weakness!

In the early days of the Internet, there were only a few people capable of hacking. Kids didn't grow up using computers every day. But today millions of kids around the world are computer experts by the time they're 18. Most of them aren't hackers or criminals, but the pool of people who are capable of hacking computer systems is enormous, especially in countries like Russia, where there seems to be very little law enforcement trying to discourage it.

It's all very frustrating, whether you are a consumer or a business trying to do online commerce. And the trends over the last few years are pretty discouraging. But the situation is not hopeless. Thousands of very smart people are working to find fixes. We will talk about some of them.

First, let's talk briefly about what we all can do right now to address this issue. Even today, with all these problems, most people are more afraid of forgetting their password than getting hacked. So they still use passwords that can be guessed. Long, random, unique passwords really do help protect you. Password managers like 1Password can also help. Another option that many online businesses and financial institutions now offer is two-step authentication. That's where you give them something you have – your password, and then they send you something, usually by text to your smartphone, that they have – a 6 digit code, that is required to complete the login process. And make sure your email account is protected by a strong password. A lot of stuff like password resets goes through email. Once hackers get into your email, they can use it to get into your other accounts!

All this stuff helps, but we still seem to be gradually losing the war. The way we prove our identities online is in need of a major upgrade. Businesses need to be able to authenticate devices and users without the need for passwords. Getting passwords and human intervention in general out of the authentication process greatly reduces its vulnerability to attack. That's because no matter how much money an organization invests in security, it's all in vain if the employees and customers use passwords that are easy to crack or steal.

Blockchain technology promises to possibly solve this problem! Let's look first at what it is, and then we'll look at how it may be able to get rid of passwords.

A blockchain is an expanding list of records linked together using cryptography. It was originally conceived in 2008 as a means of creating the crypto-currency called Bitcoin. But over the next few years, people quickly saw that it had a great deal of potential for dealing with the security problems and crime that plagues the Internet today.

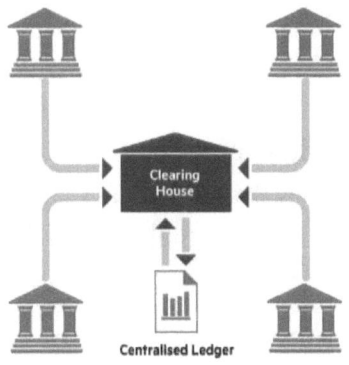

Conventional Technologies

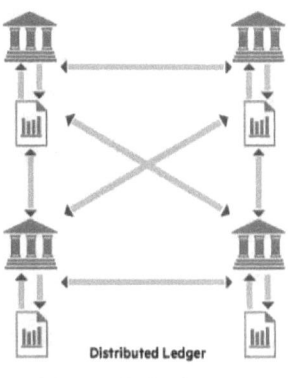

Blockchain Technologies

Blockchain works because it is inherently decentralized. It doesn't have any government authority or company maintaining its framework. Instead, it is spread out across the network in a group of nodes. Each node on the network has a copy of the chain or ledger. To add a new transaction to the ledger, each node must check its validity. If a majority of nodes agrees that it's valid, then it is added to the ledger.

Since there is no central control, there is no one who can modify a ledger entry or make any other change for their own benefit. Encryption also adds security to the system. The blockchain is maintained by a large number of users of the system. It distributes the record across a number of computers and guarantees it can't be altered.

At this point it is fair of you to ask: What does this have to do with my passwords or making my online accounts more secure? The short answer is that blockchain technology is seen as a brand new way to keep millions of strangers honest and trustworthy when interacting with each other. And there are now hundreds of companies and thousands of people around the world working to put blockchain to work solving our Internet and financial security issues.

Let's look at blockchain technology in a little more depth, and see how it improves Internet security. After that, I will give you a couple of specific examples of how it might replace your passwords.

I am not ready to say that blockchain makes hacking impossible, but it's pretty obvious that it makes it incredibly more difficult. The blockchain is distributed over many nodes – i.e. stored on many computers. Any attempt to change one copy just results in that copy getting purged from the blockchain. If you want to alter a record, you have to add a new block to the chain. It has to be signed, authenticated, and accepted by a majority of nodes. And everything is encrypted.

Blockchain really can guarantee that records aren't altered. Its structure and encryption pretty much guarantee that any attempt to alter a record will automatically invalidate it.

So blockchain technology itself is really robust, and really hard to hack. The issue is how we employ it to make the Internet, our personal data, and our financial transactions more secure.

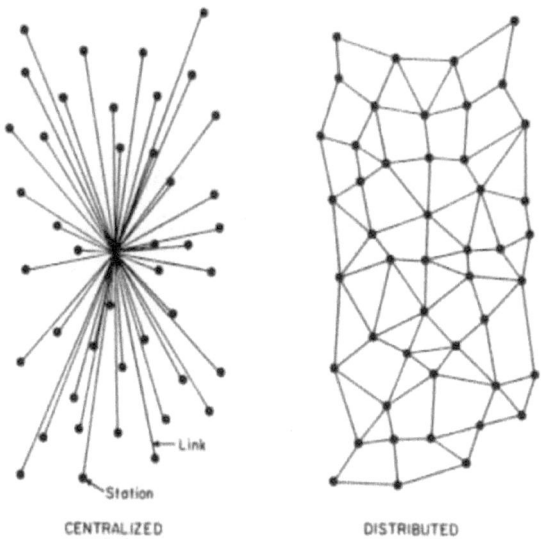

CENTRALIZED   DISTRIBUTED

Blockchain offers strong authentication, encryption, distributed data and an opportunity to do away with passwords. Some schemes for eliminating passwords using blockchain technology are already in operation, at least on an experimental or trail basis. One system is Secure Quick

Reliable Login (SQRL) which utilizes QR codes (the little square box with the swirly stuff inside) and public-key cryptography to achieve logins without passwords.

User account login

Create new account   Request new password

Username: *

[   ]

Enter your username.

Password: *

[   ]

Enter the password that accompanies your username.

[Log in]

With SQRL, there is a QR code box available as an alternative to conventional login user name and password. If the user has the SQRL app on his phone or computer, instead of entering a user name and password, he simply clicks on the QR code. The SQRL system uses cryptography and blockchain to verify his identity. It then logs him in without making him use his user name and password. While still in the experimental mode, SQRL has the promise of perhaps making online transactions much safer from hacking.

Another approach comes from a company called REMME. Its system assigns each device, like your smart phone, an SSL certificate. Certificate data is managed on a blockchain, so that fake certificates will never work. With this system, your device itself can verify your identity through its certificate, eliminating the need for passwords.

So blockchain offers a way to eliminate passwords and make the Internet much more secure. But the second thing we talked about is the need to get away from massive amounts of data stored in a single centralized location. Blockchain's decentralized nature and its hard encryption make it sound like a possible solution, but it's not quite that simple. You can't take a massive database and put it in a blockchain and then duplicate it on thousands of nodes. There are some new blockchain processes that deal with large amounts of data. A process called "sharding" distributes a large database across many nodes. The database is chopped into a whole bunch of small pieces, each of which is encrypted. These pieces or shards are stored in more than one node, but not in every one. The result is that the database is distributed and duplicated across the network. Then another process called "swarming" allows a system of distributed nodes to store and manage the data, all within the secure and encrypted technology of blockchain. But even sharding and swarming don't allow blockchain to replace big data centers. Blockchain will make them safer and harder to hack, but we are stuck with big data centers until some new technologies come along.

Will blockchain and other new technologies allow us to solve our Internet security issues? We'll have to wait and see, but a couple of things are clear. There are a lot a people working on Internet security. Things are going to change, and hopefully improve. And we are going to hear a lot more about blockchain technology. Blockchain technology is still relatively new, so we don't have all the answers yet, but blockchain clearly has the potential to have a big impact on the data security.

Fun with Digital Electronics

In this chapter, I am going to tell you about some of my hobby projects in recent years. Being a computer hobbyist was a real challenge back in the 70's, as I have already discussed in previous chapters. But today, there are hundreds of products that make it fun and relatively easy. One big reason is that single chip microcontrollers today are powerful computer systems, complete with build-in RAM memory, timers, serial ports, and flash memory for program storage. Not only that, they have these very cool integrated development environments or IDEs. IDEs are software tools that make programming microcontrollers easy using high level languages like C+, Java, or Python.

I first noticed these new microcontroller boards for hobbyists back in the early 1990s, when Parallax Corp. introduced the Basic Stamp in 1992.

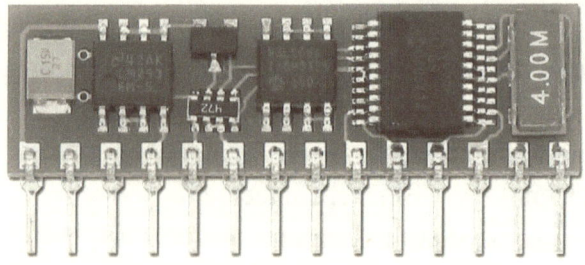

Basic Stamp

The Basic Stamp was a very basic microcontroller with 8 I/O ports, a small amount of RAM, and a small amount of flash memory, but it came with a built-in BASIC interpreter, and could be programmed from any PC running DOS. I wasn't that impressed by either its features or its performance, but I was really impressed by how easy it was to learn about and program. You could literally have it executing a program of your own creation within an hour of getting your hands on one.

A couple of other products came out more recently that have created a whole new generation of computer hobbyists.

The Arduino family of microcontroller boards started out as an open-source project in Italy in 2003, but it has become a huge global phenomenon in the last few years. Arduino has dozens of processor boards, hundreds of functional add-on boards, which they call "shields", and a powerful IDE development system. Dozens of companies make Arduino accessories, Arduino clones, sensors, motor controllers, etc. Arduinos, for the most part, use 8 bit Atmel microcontrollers running at 16 MHZ. But some boards have faster or higher performance processors. The Arduino IDE makes programming these controllers extremely easy using the C+ language.

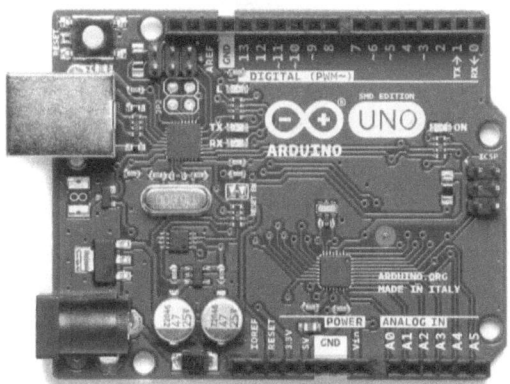

Arduino Uno

The Arduino UNO above is typical and is also the most popular Arduino board. It has 14 digital I/O pins, 6 analog input pins, 2K RAM, 32K flash memory, and runs at 16 MHZ. It costs about $20. I will talk shortly about some of my own Arduino projects.

Another product that has captured the imagination of many computer hobbyists in recent years is the Raspberry Pi. This is not a microcontroller but a full-fledged high performance computer.

Raspberry Pi

The Raspberry Pi uses a Broadcom 64-bit ARM processor running at speeds up to 1.4 GHZ. It has 1 GB of RAM, 40 I/O pins, built in video, Ethernet, WI-FI, Bluetooth, and runs the Linux operating system. It can be programmed using a variety of languages, though Python seems to be a favorite with Pi users. It costs about $35.

With the Pi's awesome amount of computer power for $35, you might wonder why I wouldn't always choose to use it rather than the slower, less powerful Arduino. The answer is that the Pi is a lot harder to use for many simple projects. Linux is a full-blown multi-tasking operating system that has a lot going on at once. Your application has to share the processor with all kinds of other functions running in the background. Critical timing is unpredictable with multi-tasking systems.

My personal choice was to use Arduino for most projects, though I did dabble for a while with the Raspberry Pi. My son-in-law, Kyle, on the other hand, is much more comfortable with Linux than I am. He chose the Pi for a bunch of projects that could have been accomplished with Arduino. So which is better is sometimes just personal choice.

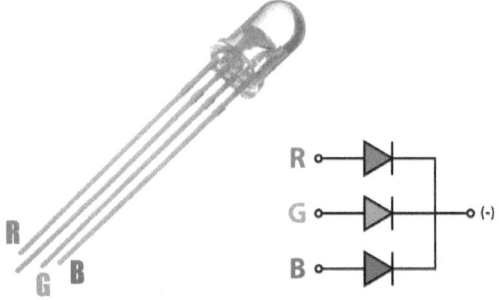

A multi-color (Red, Green, and Blue) LED is actually 3 LEDs mounted in one package. The one shown here has a common cathode, and then three anode leads, one for each LED.

Now I will talk about a couple of my hobby projects in recent years. They both involve large numbers of multi-color LEDs.

The first is a flat panel display. I started out making a small display 8 LEDs high and 40 LEDs wide. That's 320 LEDs, but each one contains a red, green, and blue LED, so there are really 960 different LEDs to be controlled. The interesting question is how do you control 960 different LEDs all at once, and update whether they are on or off about 50 times per second? That rate is required for scrolling text or any kind of animation. I will try to give you a brief answer to this question without getting overly detailed.

The processor can't control that many LEDs at once without a little help. It comes in the form of shift registers, digital ICs which can briefly store a stream of data put out serially from one of the processors I/O lines. Our shift register chains are 40 bits long, to accommodate the 40 LEDs across. We have three shift register chains: one for red, one for blue, and one for green. We load into the shift registers the data for the on/off status of the top row of LEDs in our display. We do it for each of the 3 colors. The outputs of the shift registers are tied to the anodes of all the LEDs in their respective column. With the data loaded into the shift registers, we use a new I/O pin to select the top row of LEDs and connect their cathodes to ground. At that point the top row of LED is turned on as necessary. We wait one millisecond, and turn them off. Then we quickly repeat the process for row 2, then row 3, etc. through row 8. If we can complete this whole process in less than 20 milliseconds, we can update our panel 50 times / sec. That's fast enough that the human eye sees the whole panel lit up at once, even though only 1/8 of it was actually on at any one time.

Above is my first panel

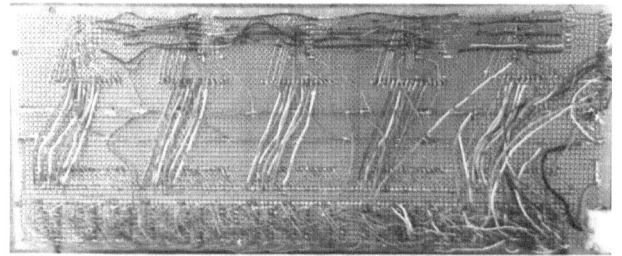

Above is the backside. The maze of wires
at the bottom is the shift registers.

As small as this first LED flat panel was, it turned out to be a lot of work because I wired up all those shift registers and LEDs by hand. That turned out to be many hours of work.

From a programming standpoint, there are a couple of interesting challenges to these displays. The first is what we just talked about – getting all those LEDs refreshed multiple times per second. It turns out loading those shift registers fast enough takes some programming tricks. The finished subroutine that performs this refresh process fast enough is a major accomplishment in itself, before you have anything to actually display on your panel. The other programming challenge that is interesting is displaying text and making it scroll. You need a font that is appropriate for an LED display. You need to store that font in your program. You need to convert a string of text into data generated by that font that can be loaded into the display, and finally, you need a system to make it all scroll across the display. All that turns out to be a fair amount of effort, but eventually it all worked.

My 40 by 8 LED display was a huge amount of work, but the end result was puny. I wanted something bigger, but trying to hand wire something bigger was out of the question. Fortunately, there are commercial LED flat panels that come with the required shift registers already built in.

My 64 x 64 LED panel

My next LED display consisted of two purchased 32x64 LED panels arranged as a 64x64 panel. It had a lot more LEDs, but was only 2 colors: red and green. So that was 8192 LEDs that had to be updated every 20 milliseconds. It turns out this is still within the capabilities of 16 MHZ Arduino processor, so this project was very similar to the smaller one.

I was using the Arduino Mega for these displays, because it has lots of I/O ports and 8K bytes of RAM. That's 4 times the RAM of the Arduino UNO. For this 64 x 64 panel, I used 4096 bytes of RAM to store the status of each LED, e.g. a byte set to 0 was off, set to 1 was red, set to 2 was green, and set to 3 was yellow (red and green both on). That used half of the Mega's RAM, and left the other half for all other programming needs.

I could do a lot more with this 64 x 64 matrix, and was able to create little animated shows for Halloween and Christmas on it. However, it was still too small. The 64 x 64 LEDs formed a display about 1 foot x 1 foot. It was still pretty puny!

I then found some new commercial panels. They were also 32 x 64 two color LEDs. But they were physically twice as big as my current panels. I got 4 of them and put them together to form a 64 high and 128 across panel that was approximately 2 feet high and 4 feet across.

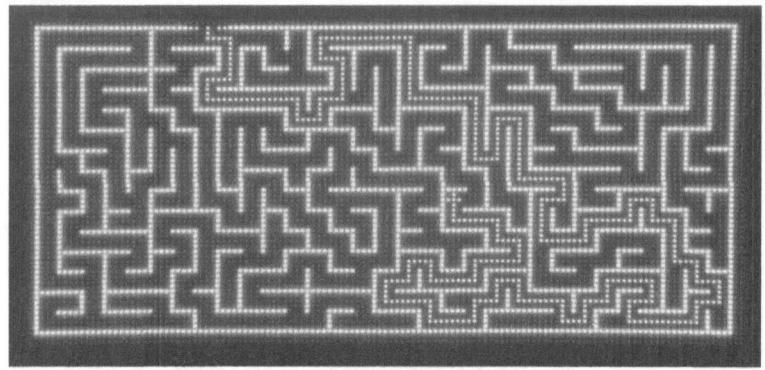

This panel was finally big enough to be useful. As shown above, I wrote a program that would create a maze, solve that maze, and then show you the solution. I could also put a Christmas display on it that is visible in my front yard.

But this new panel posed some interesting challenges for the microcontroller. There are 16384 LEDs to be controlled, or 8192 two color LEDs. My previous scheme for storing the status of each LED wouldn't work, as it used up the entire RAM just for the LED status array. I had to store the status of multiple LEDs in each byte of RAM. That's not hard to do, but it slows things down a lot when you're refreshing the panel. And with 16384 LEDs to refresh, we need as much speed as possible. Refreshing this display pushed the Mega to its maximum limits. I spent several hours fine tuning the refresh routine to maximize speed and performance. It took the full 20 milliseconds to refresh the display, without leaving any processor time to do anything else, like scrolling text or changing the content on the panel. Automatically calling the refresh routine every 20 milliseconds or even every 40 milliseconds created huge timing issues. I had to forget about timed refreshes. I set up a pattern or picture to display and then call the refresh routine, then made a change to the picture and refresh again. In other words, the program had to explicitly request a refresh, rather than have a timer do it automatically. This approach is not very elegant from a programming standpoint, but it did allow the Mega to control this big panel.

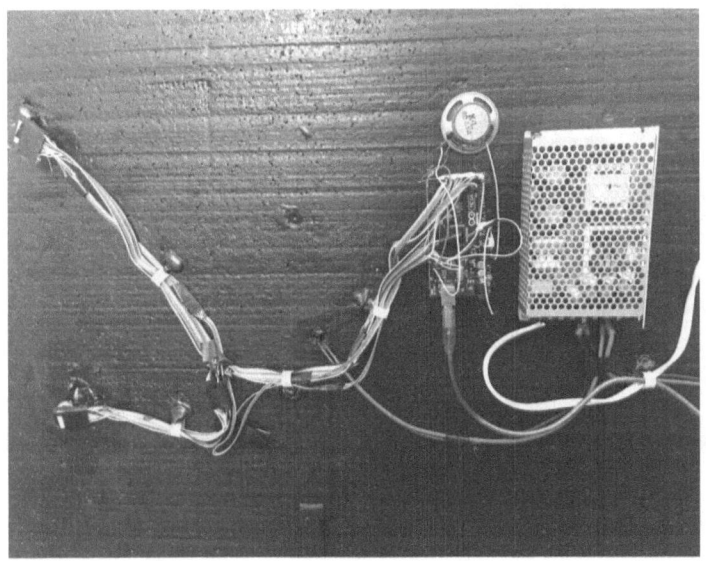

This is the back of the display. You can see the Mega, nearly covered with wires, in the center. Next to it is the hefty 16 amp, 5 volt power supply needed to light up those 16384 LEDs.

The second project I want to tell you about is the RGB (red, green, blue) LED cube. It's 8 x 8 x 8 LEDs wired together into a cube.

The cube has 512 three color LEDs, so that's a total of 1536 LEDs to control. That's not too many LEDs to control, and we refresh the cube much like we did the flat panel. First we light up the top 64 LEDs, then the second layer, working our way down to the bottom layer. And again we do this very fast, so that it looks to the human eye like the entire cube is on at once.

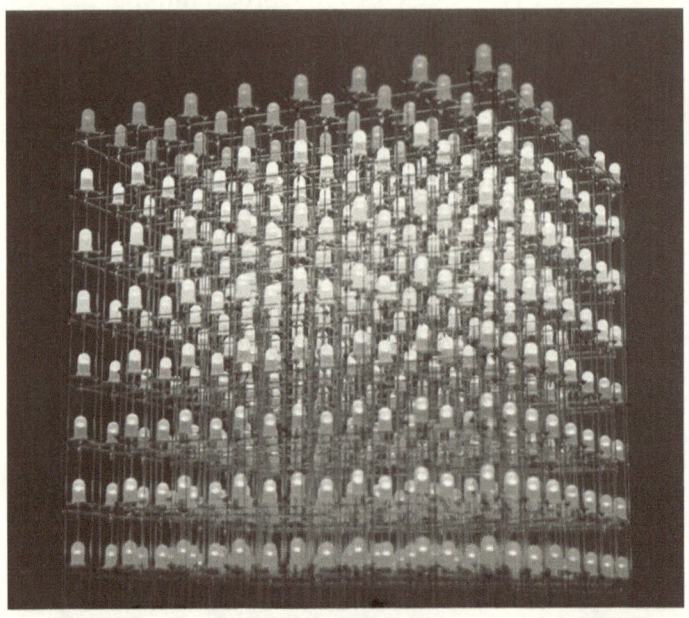

But to get interesting effects with the cube, we need to be able to generate all the colors of the rainbow, i.e. not just red, green, blue, or say orange (red and green together). We want every color. That is accomplished using something called bit-angle modulation. And it requires 3 bytes of data for each LED, indicating the intensity of the red component in one byte, the intensity of the green component in another byte, and the intensity of the blue component in another byte. The refresh cycle takes much more time when bit-angle modulation is used, so the Arduino is only barely up to the task. So I control my cube with an Arduino like microcontroller called chipKit.

The chipKit uC32 is a 32 bit processor instead of 8 bit like Arduino. Even more important, it runs at 80 MHZ instead of 16. So the chipKit runs at least 5 times faster than the Arduino and can easily handle bit-angle modulation.

The real challenge with LED cubes, especially the RGB multi-color ones like mine, is the time involved to build one. They involve building special jigs, then 80 to 120 hours of hand soldering LEDs together into the actual cube. Then there is a maze of wires coming out of the cube into shift registers in order to control it. Fortunately, there's a guy in Toronto who sells a printed circuit board that holds the shift registers and drivers required to power the cube, but the job is still enormous. That's the reason you don't see LED cubes very often.

For me, building the cube was drudgery, but I really enjoyed programming it. That makes me the opposite of most cube builders, who liked building one, but wanted to use someone else's software to make it do fun and interesting stuff.

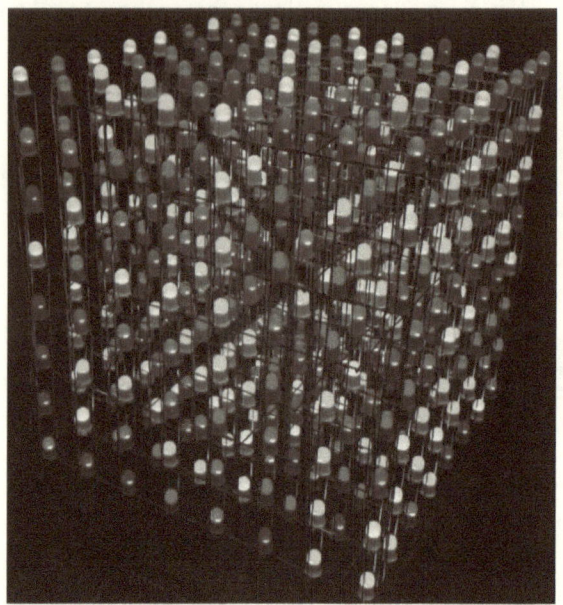

At first, I tried other people's software, but the effects were limited. I wanted to create my own. I started out completely from scratch, creating a very simple program to handle refreshing the cube, dealing with bit angle modulation, and then I created a single demo effect. It looked like a wild mouse running around in the cube, with a fading trail behind it.

I published it to a list of fellow cube enthusiasts, encouraging them to use my software platform to create their own effects. I got a lot of good feedback, but confirmed what I had already been told by my friend (the printed circuit board guy) in Toronto: everybody wants more software, but most cube owners either don't want to or don't know how to program them.

So I started to make enhancements to my software. Over the next year, I produced and published a series of ever more complicated software platforms, with dozens of special effects for the cube. For example, one iteration of my software added the ability to drive the cube from a music source, turning the cube into a color organ. Another added the ability to put a pattern into the cube and then rotate it in any direction.

Among cube enthusiasts, I apparently became known as one of only a few cube software gurus. I wasn't really trying to accomplish that. It just sort of happened. The final result was a package of cube software with about 3 dozen effects, which took about 25 minutes to watch from start to finish. It has now been used by hundreds of cube builders.

Current Computer Trends

There are a lot of subjects we can potentially cover in this chapter. I will try first to talk about near term trends, like the continuing trend toward mobile devices and the emergence of cloud computing. Then we can talk about longer term, more speculative things, like neural networks, nanocomputers and quantum computers.

It seems like everyone has a smartphone today. But actually, it wasn't until 2013 that 50% of Americans owned a smartphone. Today in 2019, that number is 80%, with an additional 15% that have a cell phone but not a smartphone. So we are getting close to the point where everyone has a smartphone.

But now let's look at computers, excluding smartphones, but including desktops, laptops, and tablets. Desktop computers today only account for about 20% of computer sales, and their numbers are slowly declining. Laptop and notebook computers, excluding the hybrid detachable tablet/ laptop, are slowly growing. Regular tablets are slowly declining. But the hybrid detachable tablet/laptops are growing rapidly. They had 5% of the market in 2016, but are expected to have 15% by 2021. They are the only piece of the business that is really growing right now. As mentioned in previous chapters, Microsoft Surface devices are the best known of these hybrid tablet/laptops, but there are other brands as well.

Microsoft Surface

It is safe to conclude that the trend toward mobile computing will continue for the foreseeable future, both in smartphones and computers.

Let's now change subjects and talk about cloud computing. It has some history that we really haven't talked about yet. Back in the 1950s and 60s, all computers were big mainframes that used batch processing. Computers were big and very expensive. It was important to keep them busy and productive, so jobs were submitted and placed in a queue. You never knew exactly when your "job" would run, but eventually its turn would come. The most common way this worked was to submit your job by some deadline, say on Monday, and your results would be available sometime on Tuesday morning.

In the 1970s, timesharing came along. It was a different approach to keeping an expensive, big computer busy. It relied on multi-tasking operating systems that attempted to handle hundreds or even thousands of users simultaneously. That was possible because an individual user didn't put much demand on the computer, and when he did need the full attention of the computer, it might be only for a few fractions of a second while it executed his program.

In the 1980s, the personal computer took over. Everyone had his own dedicated computer, and time sharing disappeared. However, that was before the Internet. As we talked about in an earlier chapter, computers became a lot more useful, and a lot more an integral part of almost all human activity, once the Internet came along.

What has happened over the last 25 years as the Internet has grown and become the center of activity for all personal, business, government, and educational activities? We have found more and more Internet functionality that relies on giant banks of servers, like the ones maintained by Google, Facebook, and Amazon. These server farms are huge, expensive facilities costing hundreds of millions of dollars. They are finding more and more applications, and all of a sudden – timesharing is back! But now we call it cloud computing.

Cloud computing is a type of timesharing of computer resources, but it's a lot more than the timesharing back in the 1970's. Cloud computing in most cases involves computer services that didn't even exist before the Internet, things like web servers, data centers, virtual computers, remote data backup, and a wide variety of Internet based services. Companies choose to use cloud computing services because they are much more cost effective than trying to provide all the needed services internally.

To understand this a little better, let's imagine a company with

1000 customers. It maintains all the data on these customers and allows them to enter their orders, make payments, etc. using two servers, each backing up the other. The servers are pretty busy during the day, but are generally idle at night. Your customer base grows to 2000, and you buy two more servers. But two years later, you have 2 million customers. Do you buy 4000 servers that sit idle half the time? It is simply much more cost effective to let a cloud computing service deal with how many servers are needed and how to keep them busy and reliably running 24/7.

Google at The Dalles, Oregon

Another reason cloud computing is cost effective is that there are many economies of scale when it comes to servers and computer services. To see what we mean, let's look at just one of Google's data centers – the one at The Dalles on the Columbia River in Oregon. There it has a couple of warehouses each holding 75,000 servers. They are at The Dalles for a few reasons. One is a large hydroelectric plant nearby on the Columbia River that supplies lots of cheap electric energy. Another is that water from the river can be used to cool the building where all that power is being consumed. Another is that the main fiber-optic cable linking

North America and Asia comes into the U.S. just a few miles to the west. Google has data centers like this around the world. This is big business on a grand scale. Very few companies are big enough to compete with the big boys like Google and Amazon. This is why cloud computing services are more cost effective for most companies than trying to do all this stuff for themselves.

Thousands of Google servers

Cloud computing systems are big business and are here to stay for a while. Their functions may change in the future, if, for example, block chain technology starts to replace centralized datacenters like Google's. But block chain too requires big, expensive computer resources, which likely will also require cloud computing.

Now let's change subjects. Let's talk about GPUs or graphics processing units. You have one in your computer and another one in your smartphone. They are the chips or chip sets that create the color images on your display. They have been evolving and getting more and more powerful over the last 30 years or so. We are talking about them here because for many specific applications, they turn out to be much more powerful computers than conventional CPUs!

The reason GPUs are so powerful and interesting is that they have massively parallel processing capability. They have always been very complex systems, but up until recently they were hardwired specifically to display graphics – they couldn't do anything else. But graphics processing technology has evolved and now much of the hardwired functions have been replaced by general purpose processors.

NVidia GeForce 8800 graphics card

NVidia's GeForce 8800 was introduced back in 2007, but illustrates how these graphics processors have become general purpose powerhouses. The 8800 has 128 processor cores each capable of performing general purpose floating-point math. The raw computational horsepower of a single 8800 chip is staggering: It achieves a sustained 330 billion floating-point operations per second.

As these GPUs have become more programmable, and more and more powerful, a new industry is starting to form around them – general purpose computation on graphics hardware or GPGPU.

GPU's can't be used for most computer tasks, but any time a problem lends itself to massively parallel processing, GPUs can run circles around conventional CPUs. Examples where GPUs are being used today include machine learning, cryptocurrency mining, blockchain processing in general, signal and video processing, and neural network computing, which we are going to talk about next.

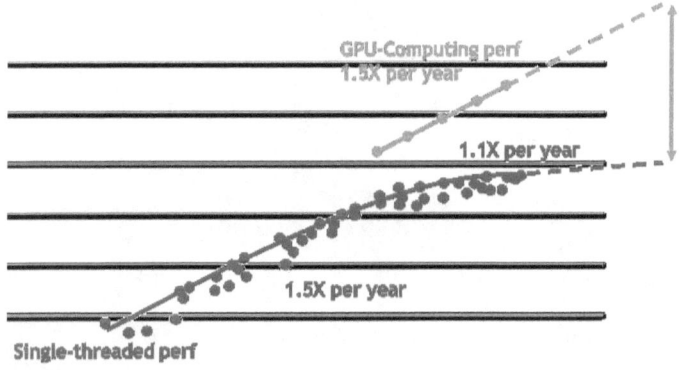

In the graph above, NVidia points out the single core processing performance improvement has slowed in recent years to about 10%/year. Where parallel processing can be employed, GPU performance exceeds CPUs, and is improving at a rate of about 50%/year! So perhaps we are going to see a lot more of GPUs in the future.

Now let's change subjects again, and talk about some new emerging technologies which don't really play a big role in digital electronics today, but may very well do so in the near future.

One of these technologies is artificial neural network computing. These systems can be modelled on today's computers or can, on a limited scale, be physically created in electronic hardware. In either case, they attempt to learn and solve problems the way animal and human brains do, using networks of neurons. The concept is very different than the way conventional computers go about solving problems.

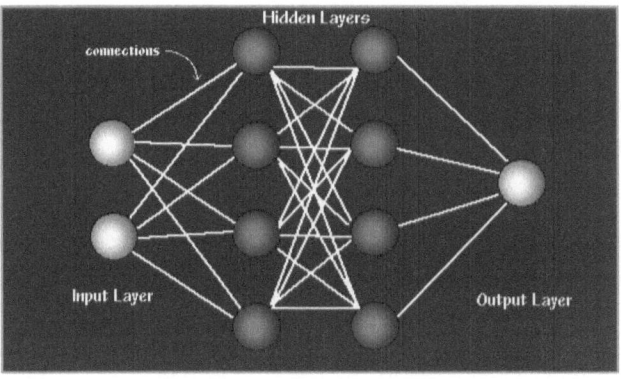

Neural networks take inputs through a series of layers where complex interconnections weigh and analyze the inputs, eventually producing meaningful outputs. Artificial neural networks typically contain a few thousand neurons, while the human brain contains approximately 100 billion neurons. But neural networks of even a few thousand neurons can produce surprisingly impressive results with tasks like self-learning or adaptive behavior based on experience.

Image recognition is an example where neural networks are clearly superior to conventional computers. You can show a neural network a thousand pictures, some of a cat and some without a cat. When there is a cat, you tell it "cat". The network learns from this experience and pretty soon it recognizes cats without being told. Without knowing anything about cats, it learned to recognize one from experience. This captures the essence of what neural networks do better than conventional computers, whether they are artificial or biological.

Today, neural networks are starting to be used in actual applications, almost always as simulations on a conventional computer. Here is a sampling of the kinds of things they are doing:

- Quality control – a neural network can inspect parts on a factory floor and identify defective ones.

- Aircraft Autopilot – a neural network can monitor the planes instruments and take corrective action to keep the plane level and on course.

- Security – a neural network can monitor thousands of credit card transactions and flag ones that appear fraudulent.

Artificial neural networks are part of the broader technology of artificial intelligence. The term artificial intelligence or AI refers to the branch of computer science dealing with the simulation of intelligent behavior using computers. It is a huge subject with huge amount of activity today. It's hard to pin down exactly what constitutes artificial intelligence, because it usually refers to leading-edge stuff. So today self-driving cars, face recognition, Apple's Siri and Amazon's Echo, and playing Chess at a grandmaster level are considered artificial intelligence, but speech synthesis and playing tic-tac-toe probably don't qualify. Our definition keeps changing as the technology progresses.

We won't go into a lot of detail about artificial intelligence, but we should point out a couple of things. Almost all artificial intelligence today is what is called weak or narrow AI – it's very good at performing a specific task, like driving a car or playing chess, but that's all it can do. The other form of AI is called strong or general. For strong AI, a computer would have to perform a wide variety of cognitive tasks. It would have to be capable of dealing with new tasks presented to it that it had never seen before. Today in 2019, we have many great examples of weak AI, but have a long way to go to achieve strong AI.

Another futuristic area to look at is nanocomputers. Or perhaps we should look at the somewhat broader topic of nanoelectronics in general. This subject starts with the continuation of Moore's Law which we have already discussed in detail.

Current leading edge technology in 2019 for manufacturing integrated circuits uses 14 nanometer gate widths, where a nanometer is a billionth of a meter. There are plans in place to continue making ICs smaller, down to 5 nanometers in the next few years. As we get smaller than that, we get into what is generally referred to as nanoelectronics, where things get really interesting.

Some people predict that Moore's Law will fail at this point, primarily because we are starting to get into an area where quantum mechanics start causing errors, i.e. electrons start to act like they are in more than one place at the same time. Others simply suggest that the technology will need to change significantly and we will find new ways to make Moore's Law keep going.

I'll just briefly mention some of the potential technologies, and then we'll talk about the potential of nanocomputers. There are things called nanowires, which offer all kinds of new properties with which to construct electronic circuitry. There is the possibility of employing biological DNA as an extremely compact computer memory technology. There are schemes under development to build electronic elements at the molecular lever using a method called molecular self-assembly. There is a form of computer memory called single electron memory, which would perhaps provide the ultimate in memory density, but it is still a long way off. There are things called quantum dots that have optoelectronic properties that could prove very useful.

All of this means computers are going to continue getting smaller, or that more and more computer power will be able to be stuffed into the same space. We can only begin to speculate on the implications of nanocomputers. Just like the early computer that filled a small building can now sit in the palm of your hand, nanocomputers might mean that a huge data center like the one Google has in The Dalles might someday fit in your hand.

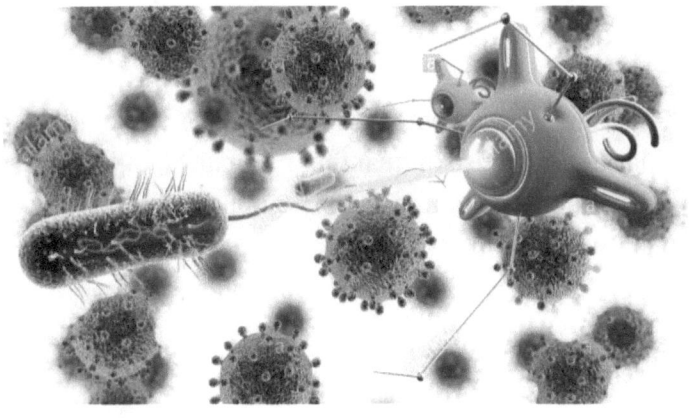

Nanobots find and destroy cancer cells in the body.

And nanocomputers could mean that we could send a computer along with a drug into your body. That computer could make sure the drug is delivered where needed and released only as needed. Nanocomputers might mean that we can build computers as smart as human beings, computers that become sentient or self-aware. We will talk a lot more about sentient computers in the next chapter.

Finally, let's talk about quantum computers. They are a little off subject, because this book is about digital electronics, and quantum computers are better described as physics projects that solve computer problems – not really a part of digital electronics. But we can talk briefly about them here, at least to acknowledge that they are a part of computer technology.

We will keep this simple out of necessity - my own understanding of the subject is very limited. Regular computers use bits – electrical 1s or 0s. Everything from emails to music to YouTube videos are essentially long strings of these bits - binary 1s or 0s. Quantum computers use something called qubits. A qubit is a property of a sub-atomic particle that has two states, which can represent a 1 or a 0. But unlike regular bits, because of quantum mechanics, qubits co-exist as both a 1 and a 0 simultaneously up until a measurement of that property forces it to choose, at which point it becomes a 1 or a 0. This simultaneity is called superposition, and it is at the heart of what quantum computers can do that regular computers can't.

Certain problems are essentially unsolvable with conventional computers. Take two very large prime numbers and multiply them together. It is very easy for a computer to do this and get the resulting product. But given the resulting product, if we ask a conventional computer to find the two prime numbers that produced it, the conversional computer can't do it. It's not because it's impossible, but rather there are so many possibilities to try that it might take 10 million years even for a super-computer. This difficulty finding prime factorials is the basis for all modern cryptography. But these problems are solvable with quantum computers. Many problems in molecular chemistry and other fields have similar issues. It makes them impossible to solve except with quantum computers.

IBM's 56 Qubit Quantum Computer

Very simple quantum computers have been built today and actually work. They are very difficult to build for a lot of reasons. One is that they must operate in extreme cryogenic conditions, typically a fraction of a degree above absolute zero. And at any temperature above absolute zero, they are prone to errors due to noise called decoherence. Elaborate schemes must be implemented to correct for decoherence errors. So the quantum computers built so far can't yet solve any problems that a regular computer can't solve, but progress is being made.

The term quantum supremacy is used to describe the point where quantum computer technology can solve problems that regular computers can't. It may take a few more years to get there, but people working on the problems seem confident that quantum supremacy will be reached. In the meantime, several companies like IBM are already offering use of existing simple quantum computers through cloud computing services!

I can't end the discussion of quantum computers, and leave you with the impression that they are soon going to ruin all prime number based cryptography. All Internet security schemes use this cryptography today. And once quantum computers get well past quantum supremacy, things are going to have to change. But experts in cryptography and quantum computers assure us we will have even better cryptography as a result of quantum computers.

What's the Future
Hold for Computers?

Now let's take a long-term view of the future of computers.
Let's ask not only about where computers are going in the next
couple of hundred years, but also where mankind is going as a
result of computers.

For the past 10,000 years, and very likely for the past 50,000
years, humans have basically had a consistent level of
intelligence. What has changed over time is the accumulation
of and access to knowledge. With the invention of the printing
press, and more recently the computer and instantaneous
global communication, the cumulative knowledge available to
us has been rising exponentially for the past 300 years.

Imagine going back in time to 1850, just a little over 150 years
ago. No electricity, no telephone let alone cell phones, no
movies let alone television, no adding machines let alone
computers, the Internet, email, etc. How would you try to
explain television, computer games, and the Internet to
someone in 1850?

We have to accept as a given that technology is changing our
world so fast today that it is hard to imagine what the world
will be like in 150 years. Human knowledge and technology
are both growing exponentially. That means that the next 150
years will see changes more unimaginable to us today, than

television, cell phones, computer games, and the Internet would have been to someone back in 1850.

So let's look at what happens 1) when we see really powerful new interfaces between computers and humans emerge, 2) when computer intelligence reaches or surpasses human intelligence.

We're not the first one to think about these things. Ray Kurzweil published a book in 1999 called "The Age of Spiritual Machines," in which he discussed the evolution of computer technology. He presents a very detailed and convincing argument that machines will be as intelligent as humans in a mere 50 years, and that humans and super intelligent machines will begin to merge into something greater than either one in a mere 100 years.

Taking this line of reasoning a little further, one can speculate about a moment in time in the not too distant future when the intelligence of man and his machines reach some "critical mass" where computers start teaching themselves and intelligence rises in a "big bang" like phenomenon to something far exceeding what we are capable of conceiving today. Ray Kurzweil discusses this event in his 2nd book "The Singularity Is Near: When Humans Transcend Biology" published in 2005.

So I will attempt to summarize here some of the Kurzweil's ideas. I will also mention some of the social and moral dilemmas that these possibilities present.

Computer technology needs only to continue progressing at its current rate of improvement for another 50 years to make computers smarter and more compact than the human brain. There is a debate about whether we can continue at the current rate (i.e. Moore's Law) for another 50 years, and there is another debate over whether we can ever make a machine truly think. Progress may slow down, and there may be some setbacks. But even if it takes 300 years, it seems inevitable given the phenomenal rate of technological progress today. At some point the power of computer technology reaches a "critical mass", where all of a sudden the whole is more than the sum of its parts, where not only does a computer attain some primitive form of consciousness and awareness, but becomes capable of increasing its own "intelligence".

Computers are already networked together much better than we humans are networked. They can exchange vast amounts of information instantaneously. As soon as you get two computers together, each as intelligent as a human, they can begin communicating in a manner that can only be described as telepathy if it were taking place between two humans. This is the nature of the singularity that Kurzweil predicts. These two computers would be instantly talking to each other and possess a combined intelligence much greater than a human. Give them a few seconds or a few minutes and they will put together a network of 100 intelligent computers with a collective intelligence 10,000 times that of a human!

A lot of science fiction and some more serious work have delved into this subject. Most of it involves the dangers and the possibility of computers taking over the world or destroying mankind. But history suggests you can't stop the advance of science and technology. We might be better off resigning ourselves to the fact that it is coming and find effective ways of making it safe and useful.

Now, changing subjects a little bit, computer science isn't the only place where technology is moving forward very rapidly. Medical science, and perhaps more specifically, genetic engineering, is on the verge of improving mankind, accelerating the natural processes of evolution. They have the potential to make us live longer, make us much smarter, and give us the capability of pushing our own knowledge and technology ever faster.

Much has already been written about genetically engineering humans, and again, most of it warns about the dangers of going down this immoral path. But even if the experts could agree that genetic manipulation of humans is too dangerous or too immoral, the real issue is whether it can be stopped. We need to assume that people somewhere, somehow will figure out how to do this stuff. And smarter, better engineered humans will be the result!

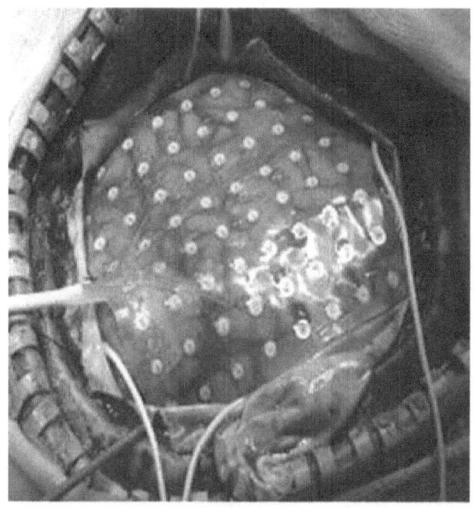

Better Computer Interfaces: This is what researchers are doing today. Imagine where we will be in another 75 years?

Now let's talk about getting improved humans and super-smart machines to work together. Even today's computers have the speed and networking capability to provide the means for human telepathy if only we could be hooked up directly to them. Medical science is going to devise effective ways to do that. A great deal of effort is already underway, because linking our brain to a computer provides solutions to today's most difficult medical problems, such as blindness and paralysis.

Medical science will continue to move forward. Computers will continue to get faster and smarter. The computer-human interface will improve dramatically. So imagine both humans and machines with higher intelligence than we have today, working together, networked together. These things are coming - they are not that far away! The results will likely take the course of mankind into realms well beyond our current capability of comprehension.

Of course, when you imagine both computers and humans with intelligence superior to our own, and further imagine seamless, instant communication between man and machine, you start to see all sorts of other options emerge. Machines have back-up memory. A computer with human intelligence, if properly designed, could, for all practical purposes, be viewed as an immortal. And if the man-machine interface is good enough, man himself could easily become a candidate for immortality!

Human kind is now on the brink of this cataclysmic change, as a combination of medical science and computer technology are now poised to significantly increase human intelligence! We are moving into a new age, one where technology is not just improving the gadgets with which we surround ourselves - it is about to start significantly changing who and what we are!

By the end of this century, we will have technology at our disposal that will create unimaginable challenges to our concepts of life, death, God, and everything else we've believed for the past 2000 years. Anything attempting to do that will, of course, be met by vigorous opposition. After all, who are we to play God? If you think cloning, abortion and stem cell research are major moral and ethical questions, you haven't seen anything yet. These are merely the first ripples of a huge storm of moral and ethical dilemmas on the horizon.

We will see resistance to intelligent machines, first from a concern that they may become very dangerous, and later, because they violate our sense of being unique or special, at least here on Earth. Finally, when they inevitably become smarter than their biological counterparts, they will be resisted as would alien invaders, not only because they are perhaps dangerous, but also because they are envied for their superior capabilities. If we maintain control of the technology, this will remain a battle between two groups of people who disagree about how smart we should allow machines to become. If we lose control, this will be a war of survival between man and machine – a real life version of The Terminator movie series!

We will also see a different kind of resistance to the merging of man and machine. Here, the issue is not man being violated by another species, but rather man's willful decision to let himself become something entirely different than what he started out to be. When people are presented with the opportunity to evolve their consciousness into a machine, increasing their intelligence and giving them immortality, some will jump at the chance and some view it as the destruction of mankind. This will be a major battle fought between two groups of people who likely will never understand each other's point of view.

We can only speculate over the outcome of these battles. But history suggests these battles will only briefly slow the progress of technology. Something that might otherwise happen in 75 years, may take 125 years. Still not very far away! As the saying goes, "once the genie is out of the bottle."

So this battle will be between a coalition of forces aligned against the reinvention of man versus the irresistible force of the onward march of technology itself. The technology may be slowed down. It may be driven underground. But it won't be stopped!

Another area of speculation might be how the losers fare. Though I keep saying technology won't be stopped, I don't mean to imply that the human race is automatically destined to obsolescence or destruction by its own inventions. Is it possible that people wanting to remain faithful to their genetic roots, who are willing to give up higher intelligence and immortality to keep their flesh and blood body, can coexist on this planet with machines that have evolved from humans? Perhaps that's possible for a few years - perhaps forever. Perhaps the higher intelligence we keep talking about will include a tolerance for others that man has never been able to achieve.

Final Thoughts

We've covered a lot of both history and stories regarding digital electronics. I tried to focus on the second half of the 20th century, but the stuff going on today in 2019 is pretty interesting as well. One thing is for sure – the story of digital electronics is just getting started.

It would be difficult to explain the Internet and a smartphone to someone in 1900, and there will be all kinds of developments over the next 100 years that we can't even imagine right now!

There are so many emerging digital technologies. We haven't talked about many of them at all yet. Self-driving vehicles are going to significantly change the way we live.

Human-like robots will also significantly change our lives. One of their first places to make a big difference will be in providing companionship and care for the elderly.

The Internet of Things is another area we haven't talked about. What happens when all our appliances are smart and can communicate with each other? Refrigerators are starting to order our groceries for us even now.

And digital electronics is going to have a huge impact on health and the healthcare system. Sensors in your clothes or smartwatch may soon alert you to an approaching health issue long before your own body decides to tell you. And that tricorder we all saw on Star Trek may soon be a reality – at least as a diagnostic tool for physicians. In fact, in 2017, Qualcomm and the X Prize foundation awarded $2.6 M to two engineers turned medical doctors who developed a diagnostic tricorder! (I can't see it mending broken bones or repairing a ruptured appendix anytime soon however.)

Even if we let our imagination go wild, I doubt we can begin to foresee all the amazing things digital electronics will bring up over the next fifty or one hundred years.

So many of the major innovations in digital electronics can trace their origins to the early 1970s, when semiconductor technology reached the point where thousands of transistors could be placed on a single chip. But even visionaries like Alan Kay back then couldn't imagine today's smartphones. It just hadn't sunk in yet what hundreds of millions of transistors on a single little chip of silicon might make possible.

As Moore's Law continues to shrink transistors, we have a few ideas about what is coming – computer consciousness and robots that can interact with us as though human are good examples. We're not talking about Siri or Alexa here – they are pretty primitive and actually do a pretty poor job at trying to act human. However, that's all going to change in the not too distant future. But what else is coming – I admit, I don't know, and I doubt anyone does.

It might be reasonable to ask at this point who deserves the biggest credit for bringing us digital electronics. We've mentioned many of the famous names like Robert Noyce, Gordon Moore, Steve Jobs, and Bill Gates. These men do deserve special mention in this book, but the truth is, today's digital electronics are the result of the efforts of hundreds of thousands of hard working fab processing engineers, process development specialists, IC designers, applications engineers, systems engineering specialists, computer scientists, software developers, manufacturing equipment engineers, computer and digital technology hobbyists, and many other technologists that I have failed to mention. I have had the privilege of working with people in every one of these fields during my employment at Fairchild and Motorola. The accomplishments of even the most famous people in semiconductor and computer technologies are relatively small compared with the collective accomplishments of so many people!

Index